Stem Cell

A Guide to Stem Cell Therapy

(How Stem Cells Are Disrupting Medicine and Transforming Lives)

Richard Herzog

Published By **Jordan Levy**

Richard Herzog

Stem Cell: A Guide to Stem Cell Therapy (How Stem Cells Are Disrupting Medicine and Transforming Lives)

ISBN 978-1-998038-80-0

Legal & Disclaimer

The information contained in this book is not designed to replace or take the place of any form of medicine or professional medical advice. The information in this book has been provided for educational & entertainment purposes only.

The information contained in this book has been compiled from sources deemed reliable, and it is accurate to the best of the Author's knowledge; however, the Author cannot guarantee its accuracy and validity and cannot be held liable for any errors or omissions. Changes are periodically made to this book. You must consult your doctor or get professional medical advice before using any of the suggested remedies, techniques, or information in this book.

Table Of Contents

Chapter 1: Stem Cell Therapy

Stem cells are truly some of the most fantastic cells within the human frame. These are undifferentiated cells that do not have an instantaneous "blueprint" or specific future. The can come to be differentiated into specialized cells anywhere inside the direction of the body.

They are classified as precise kinds of cells, those which is probably from embryonic beginning and people called grownup stem cells.

In the developing embryo, the ones cells differentiate into ectoderm, endoderm and mesoderm. These provide upward push to our spine, nerves, and all our organs. Adult stem cells are often used to repair, top off and regenerate tissues.

Historically, stem cells can come from pretty some tissues. These encompass umbilical wire, fetal tissue, bone marrow, or the outstanding belongings which is probably adipose or fat cells.

Adipose-derived stem cells have the best numbers of cells while gathered and tested compared to all others. This is through a long way the popular technique of stem cellular treatment because of sheer numbers, and the reality that they are coming out of your non-public body. This is referred to as autologous remedy.

Stem mobile research in this u . S . Has been in life for over 60 years. There are a great variety of research and articles describing its dramatic gain for persistent diseases. Many of those publications are to be had that permits you to take a look at on my net website.

In appearing stem cell treatment, fantastically strict suggestions need to be discovered in coordination with a specialized scientific trial evaluation board. This guarantees accuracy, sterility, and top notch manipulate of the system. This information collected from the manner, on the side of various types of documentation, may be used for medical ebook at a later date.

Physician notes and method, as well as a questionnaire stuffed out thru patients periodically, are part of this tool. This makes the satisfactory level of device and documentation possible.

The following instances are real patients seen in my workplace, that underwent stem cell remedy for an entire lot of issues. Of direction, I am now not able to use names or any non-public information from the sufferers, but those are documented instances and documented comments straight away from sufferers or from circle of relatives human beings as to their response and improvement.

All the patients that I deal with with stem cells file their signs and signs and signs and symptoms and issues before remedy of their very very own phrases and their non-public handwriting. Their enhancements are documented and these are a part of our chart document.

All of these sufferers might be observed up yr to year, to document development and results, that is all part of the scientific trial protocol.

Case 1: Alzheimer's Disease

This is an 80-one year-vintage female affected man or woman diagnosed with Alzheimer's disorder with the aid of her neurologist. She became treated with stem cellular treatment in August, 2014. During workplace visits the patient did not communicate a bargain. According to her son and daughter-in-law, she changed into in a steep decline in advance than coming to the place of business.

After stem mobile treatment changed into completed, her son wrote an outline of her improvement. The affected person had a stem cellular treatment achieved within the morning and on the identical day there has been a sizable improvement because the family become riding domestic within the automobile on August 21, 2014.

She engaged in non-stop communique all the manner lower back domestic. They stop for dinner and the affected person opened and observe the menu and determined upon consuming fish. She have been not able to observe a menu or use utensils for years.

When the entrée arrived she became able to eat with a fork and a knife. This end up without a doubt three hours after stem cell remedy become achieved. The own family have become astonished.

The following weeks located:

•   The affected person modified into studying extra

• The affected man or woman had greater inquisitive about television

• The affected character emerge as more privy to her environment

• The affected person modified into greater willing to preserve on conversations with the own family

• The affected person have become extra agreeable and bendy

• The affected person has a much better thoughts-set

• The affected person maintains to decorate

• The own family is quite glad collectively with her development.

Case 2: Asthma /COPD

This is a seventy -365 days-vintage male affected individual with a chronic data of bronchial bronchial asthma /COPD. He had current hassle breathing, with even the easy sports over the last severa a long term. The patient become on inhalers and tablets to three instances every day, and occasional steroid remedies.

The affected person underwent stem mobile remedy in June, 2014, and had slow however innovative advantages month to month.

At sort of five to six months after stem mobile remedy, he did not use any hypersensitive reactions, nebulizers, respiratory remedies or steroids the least bit.

His respiration is absolutely clean. The affected individual is prepared to stroll miles a day,with none wheezing or any shortness of breath.

Case 3: Chronic Back Pain

This is a 70-year-antique male who had chronic once more ache for lots, many years. The affected individual had used chiropractic treatment belts and helps, similarly to severa ache medicines.

Nothing helped. He underwent stem cellular treatment in January, 2015. The affected character did intravenous similarly to hip injection.

The patient noted and documented in his own handwriting that inside weeks of doing stem cell remedy his once more ache is honestly long long past. His hand-written phrase documents this in our chart.

Case 4: Parkinson's Disease

This is a seventy three-twelve months-vintage woman who turned into recognized six years in the past with Parkinson's ailment. She come to be on various drug treatments for Parkinson's and excessive blood strain.

The affected character did stem cellular remedy in December, 2014. She had intravenous and nasal software program program.

Within the only-month comply with-up, the affected character defined that she's doing a whole lot higher. She's walking a amazing deal greater, she described that her toes don't stay with the floor as in advance than.

The affected man or woman can open bottles now, despite the fact that she couldn't in advance than. She's feeling extra strength in her shoulders. She's able to do greater house duties and errands.

The affected individual feels happier, and has much less within the way of tremors, and plenty a whole lot less stiffness. She started going lower lower returned to the fitness center and modified into able to begin doing some strolling. She said that she no longer reports foot drop.

Case 5: Diabetes

This case is a forty-year-antique male who has been diabetic for over seven years. He is on oral medicinal pills further to Victoza njections. His fasting blood sugars have been over 250 - three hundred.

Stem cellular treatment have become performed in late July, 2015, and progressively, over three weeks, the affected individual's blood sugar declined with the resource of a hundred points. The affected person had no longer began workout or the low-carb eating regimen as however.

I definitely have reduced his medicinal drug dosage therefore and assume him to be off treatment in the very near future.

Case 6: Shoulder Pain

This affected person is a fifty five-three hundred and sixty 5 days-antique girl with intense left-shoulder ache during the last 5 to six years. The MRI found arthritis. She turn out to be now not able to do any sports activities activities or most of her every day sports activities sports. She had intense ache looking to placed on a jacket.

Stem cell remedy changed into performed October, 2014, and internal 3 weeks the affected person stated that her pain had decreased through half of.

By the second one and zero.33 month, the affected character stated being capable of adventure her bike and had notable type of movement. She had no pain accomplishing for gadgets and changed into able to sleep without pain. She had no issues in styling her hair. Now she's able to swim and adventure her motorbike for 25 miles. She has whole type of movement.

Case 7: Osteoarthritis and Shoulder Pain

This is an 80-three hundred and sixty 5 days-antique patient with extreme shoulder ache and osteoarthritis in masses of her joints. She had muscle ache in her neck and once more, further to muscle cramps. She has continual fatigue and COPD. She has had a regular shortness of breath for the beyond few years.

Dental remedy have come to be completed in May, 2015, and over the direction of the month, the affected character describes her shoulder pain as having decreased step by step via seventy five percentage. She has skilled a great decrease in muscle aches, pains and spasms. She is now capable of preserve and do her every day sports activities with out immoderate fatigue or shortness of breath.

Case eight: Multiple Sclerosis

This is a 38-yr-antique woman with an MS analysis from 1999. The patient changed into in a wheelchair that's motorized as she changed into no longer capable to stroll. She moreover had trigeminal neuralgia and arthritis in her joints. She have been going for bodily therapy with little or no response. Stem cells were completed on July 29, 2015.

Intravenous similarly to intranasal treatment had been completed.

On August 19, 2015, the affected person provided to the place of job together along with her husband. She defined doing a long way higher. She has a large Increase in power. She's in a position to stroll at her physical treatment facility on her very own for three hundred feet!! She become not capable of do that before the stem mobile remedy. Both she and her husband are incredibly satisfied.

Case 9: Chronic COPD

This is a fifty 8-12 months-vintage female who had chronic COPD for decades. She'd been the usage of oral drug remedies and inhalers for over 10 years. She moreover had rheumatoid arthritis. She felt that her troubles had been getting worse steadily.

On August four, 2015 she underwent stem cellular remedy intravenous similarly to with a respiratory nebulizer.

She over again to the place of work for observe-up on August 25, 2015. She described that she can awaken now with out coughing. Her breathing is dramatically superior. She has now not been the usage of her inhalers. She said that she's surprised that in honestly three weeks she's had a lot improvement.

Chapter 2: The Importance Of Stem Cells

Stem cellular studies has come a long manner as it commenced in the Nineteen Fifties. Today, researchers, scientists and healers are passionate about what's been taking vicinity in the vicinity of regenerative remedy plans this is changing the direction for loads degenerative illnesses.

Because stem cells have the capacity to regenerate tissues, treatment plans for plenty sicknesses have grow to be viable for the first time.

All stem cells have those houses in common:

1) They can divide and renew themselves for lengthy intervals of time.

This is different from many specific kinds of cells, together with blood cells, muscle cells and nerve cells. Stem cells can create a populace of heaps and thousands over a duration of months.

2) They are unspecialized cells.

They do now not have tissue-specific systems but they can generate specialized cells which include blood cells, coronary coronary heart muscle cells and nerve cells.

In their unspecialized states they're able to't carry out specialised roles. They can't assist to pump blood as a heart muscle cellular need to. They can't transport oxygen molecules within the bloodstream like a purple blood cellular does.

But the specialized cells they offer upward thrust to can carry out those specialised capabilities. They can renew themselves thru cell department, even after they have been inactive for a long term.

3) They can generate specialized cells.

This device is known as differentiation. Stem cells have the capability to turn out to be particular mobile types in childhood and throughout boom. Many of them offer inner restore.

They can divide indefinitely, replenishing mind cells, muscle cells and crimson blood cells and other cells which might be greater specialised. They can become cells which might be organ- or tissue-precise with specialized skills, under positive conditions.

In bone marrow and the intestine, and in some extraordinary organs, stem cells repair and replace damaged tissues thru frequently dividing. In a few organs similar to the coronary coronary coronary heart and pancreas, stem cells can divide, however this first-rate happens at the same time as certain situations study.

Embryonic stem cells have the first-rate versatility of all stem cells.

In 1998, a way of deriving human embryonic stem cells from human embryos have become devised. Cells had been then grown in a laboratory. Initially those embryos were used for in vitro strategies for copy functions. Afterward they had been used in studies with the donor's consent.

In 2006, it became discovered that some individual cells may be delivered to a stem cell form of kingdom through genetic reprogramming. These are called pluripotent stem cells (iPSCs).

Adult stem cells are also called somatic stem cells. These are cells of the body, as adversarial to say eggs or sperm cells, or germ cells.

Adult stem cells display up in extra tissues than had been previously believed. Adult stem cells had been decided in blood vessels, bone marrow, gut, heart, liver, ovarian epithelium, peripheral blood, skeletal muscle, pores and skin, tooth and testes.

Adult hematopoietic (blood-forming) stem cells from bone marrow have been used efficaciously for 4 a few years. It is posited that these cells is probably beneficial for transplantation of various sorts.

One vital sort of stem mobile is the fat- or adipose-derived stem mobile which is also referred to as an adipocyte. In human beings who've degenerative health situations, those cells won't be released sincerely.

But they'll be used to tremendous gain as soon because the cells have been extracted, centered after which administered to human beings stricken by many degenerative illnesses.

Adipose-derived stem cells (ADSCs) are notably beneficial because of the fact they may be able to trade into new tissues. They can also suppress the pathological immune responses commonplace of autoimmune conditions.

ADSCs have been applied in treatment of autoimmune more than one sclerosis, osteoarthritis and rheumatoid arthritis.

Adipose tissue has been located to be a wealthy supply of mesenchymal stem cells. One gram of fats incorporates greater than 500 times the mesenchymal stem cells as can be discovered in one gram of bone marrow.

Mesenchymal stem cells, or stromal stem cells, deliver rise to bone cells, cartilage cells, connective tissue cells, fat cells and pancreatic islet cells. Mesenchymal stem cells are inside the pulp of little one tooth, the blood of umbilical cords, fat and muscle.

Mesenchymal stem cells are called multipotent cells. This is because they've the awesome functionality to emerge as more than one tissues. Stem cells are fertile floor for treatment of diabetes, heart sickness and different illnesses.

Stem cells have already been used to generate chondrocytes, osteoblasts and osteoclasts (cells decided in healthy bone tissue). Research on animal models trying out the functionality of in vitro cells have visible promising achievement in this region.

It's hoped that powerful treatment may additionally moreover emerge as feasible for such usage as regeneration of bone from bone marrow stroma, and development of insulin-producing cells for kind 1 diabetes, and the developing of cardiac muscle cells for the repair of broken coronary coronary heart muscle after a coronary coronary coronary heart attack.

Research is wanted to analyze extra and to plan more green strategies of generate massive numbers of person stem cells for recuperation uses.

Chapter 3: Arthritis And Joint Degeneration, And Stem Cell Therapy

There are extra than a hundred forms of arthritis. What the ones all have in common is persistent joint, bone and muscle pain, alongside facet swelling and stiffness.

There are classes of arthritis: degenerative and inflammatory.

Mesenchymal stem cells (MSCs) display off effective effects of immunosuppression and anti-infection. These cells help to bring about regeneration inside the neighborhood tissue surroundings.

MSCs are best to be used in treatment for degenerative joint ailments such as osteoarthritis and rheumatoid arthritis.

Osteoarthritis is degenerative, meaning deterioration takes region as time is going on, due to joint put on and tear. Osteoarthritis is also called degenerative arthritis and degenerative joint sickness.

There are forms of osteoarthritis.

Primary OA goals older adults and is attached with factors which can be genetic and/or familial. Secondary OA takes location due to harm, deformity or misalignment.

Osteoarthritis is the umbrella term for a collection of illnesses in which the joints degrade. This includes articular cartilage and subchondral bone.

Osteoarthritis can motive locking, stiffness and tenderness of the joints. It additionally causes joint ache, and from time to time can reason a assemble-up of joint fluid.

Cartilage can also in the end offer much less safety to the bone surfaces, leaving bone exposed and prone to turning into damaged.

The affected person will commonly have a tendency to step by step lower the quantity of hobby because of the ache. Muscles in the vicinity are liable to atrophy. Ligaments can loosen.

RA is inflammatory, the end end result of autoimmune assaults on joint lining. Swelling and inflammation are hallmarks.

Other inflammatory forms of arthritis are ankylosing spondylitis, gout, juvenile idiopathic arthritis, lupus and psoriatic arthritis.

Rheumatoid arthritis affects 10 percentage of the world's population. It reasons loss of function, ache, stiffness and swelling inside the joints, most customarily inside the arms and wrists. RA is maximum not unusual amongst older adults, even though extra younger people, which include youngsters, may want to have it.

Unlike OA, rheumatoid arthritis may want to have an effect on extra than the joints. It can hit eyes, mouth and lungs, for instance. As an autoimmune circumstance, RA is because of the frame's tissues being attacked by means of the immune device.

Arthritis most often hits humans over 60, despite the fact that any age or history can get it. Previously the most effective strategies to deal with arthritis were lots less than perfect. The goals of these techniques are to reduce ache and restriction loss of feature.

This is accomplished for some with the resource of manner of transferring a whole lot much less. For others who're capable of be greater energetic, exercise is useful.

Drug remedy (oral, at the pores and pores and skin, or injected within the joint) may be utilized in greater extreme instances. Surgery to realign, remodel or replace a joint is used within the worst of them.

The consequences of arthritis create a unhappy scenario, however stem mobile treatment offers huge recuperation.

Research on animals the usage of stem cells has demonstrated remarkable effects.

Stem cells assist boom the restoration of tissue that has been broken, and beautify the immune gadget. This permits the frame to combat the disorder, and to decrease ordinary responses.

ADSCs lessen contamination and repair damaged tissue. Stem mobile remedy can inspire recovery in articular cartilage (cartilage that strains the head of a joint).

This cartilage doesn't have nerves or blood vessels, and has a low mobile content material with few chondrocytes. It repairs itself slowly.

But stem cells manufacture chondrocytes. Chondrocytes manufacture cartilage. Cell life-style strategies for manufacturing greater chondrocytes have been improving.

ADSCs, like bone marrow MSCs, can manufacture incredible sorts of cells. ADSCs can manufacture bone, cartilage, fats cells and muscle cells. They can sell angiogenesis (blood vessel formation).

After ADSCs are remoted from fat tissue thru minimal liposuction strategies, they may be isolated and purified inside the laboratory. In preclinical trying out, a dose of ADSCs are injected into animals who have osteoarthritis.

This method can suppress the inflammatory activities, diminishing destruction of cartilage and harm to the ligaments that be a part of joints.

Stem mobile remedy can deliver healing and recuperation that within the beyond have emerge as extraordinary for human beings with arthritis.

Alzheimer's disease and Stem Cell Therapy

Alzheimer's illness brings with it a lack of memory that has an inclination to emerge as greater intense as time moves on. The character with AD can be now not succesful to research. In excessive cases, delusions, hallucinations and super psychotic signs and symptoms and signs can stand up.

Alzheimer's sickness afflicts tens of hundreds and heaps. AD comes considerable variety 6 as a cause of demise, and influences 1 in 8 human beings. It is the most commonplace neurodegenerative (inflicting lack of feature or form to the nerve cells) disease, and the most common reason of dementia.

For people over the age of 60, dementia is variety 4 a number of the global's illnesses, constant with a World Health Organization file from 2003.

Treatments were restrained, relieving symptoms however bringing no remedy. But this is changing.

Stem mobile therapy has been a fulfillment in the remedy of amyotrophic lateral sclerosis (ALS), Parkinson's illness, spinal cord harm, and stroke, among other neurological situations.

ADSCs, just like the ones derived from bone marrow and umbilical wire blood, can be transdifferentiated into nerve cells. Research done on animals propose improved cognition.

Memory and analyzing were extensively more potent in folks who received stem cellular transfusions, in comparison to folks that received the transfusion with out stem cells.

In September, 2012, research led jointly with the aid of the usage of Seoul National University Professor Yoo-Hun Suh and RNL Bio Stem Cell Technology Institute (SCTI) director Dr. Jeong-Chan Ra showed proof that adipose-derived human grownup stem cells might also save you Alzheimer's ailment.

Human adipose-derived stem cells (hADSCs) had been injected intravenously into mice who had physiology and signs and symptoms and signs and symptoms and symptoms and symptoms of AD. The aim was to reverse Alzheimer's infection in mice that had it, and to prevent its improvement in mouse models that didn't have it already.

The cells moved from the injection web web page within the tail vein, thru the blood thoughts barrier, and into the mind. Mice had been injected often, with injections taking place weeks apart from the 1/three to the 10th month.

The injections brought approximately the numbers of amyloid plaques and Aβ degrees inside the thoughts to decrease, and greater memory and studying. Even 4 months after the remedy, the ones fantastic effects had been still being seen.

The mice were better able to check and don't forget. Neuropathy lesions decreased.

It is theorized that the stem mobile injections reduced beta amyloid and APP-CT which spoil mind cells. The injections accelerated neprilysin which breaks down toxic proteins with the useful resource of a chemical response with water.

Stem mobile injections together with those appear to have the capability to prevent Alzheimer's sickness, even preserving it from emerging inside the first place. Nothing like this has ever been seen in advance than.

Several anti inflammatory and neuronal growth factors had been bestowed with the resource of stem cells. In specific, the expression of IL-10 have become improved. Apoptosis (programmed cellular demise or PCD) of mind neurons grow to be suppressed.

When this suppression is finished Alzheimer's disease is thwarted.

In the past, transplanted bone marrow-derived MSCs have been placed so one can get higher reminiscence loss in mouse models with AD. But the ones mobile types can't be used intravenously.

ADSCs, however, may be given intravenously without the threat of rejection via the immune device or forming tumors.

The waft from mouse fashions to residing human cells has been made. A March 4, 2014 article on ScienceDaily.Com referred to that pores and skin cells taken from early-onset Alzheimer's patients have been transformed into the varieties of nerve cells that were broken with the aid of manner of AD.

Chapter 4: COPD And Stem Cell Therapy

Chronic obstructive pulmonary ailment end up the 1/3 vital cause of lack of life of Americans in 2012, in line with the CDC. COPD affects greater than 2 hundred million human beings round the location.

More than 3 million human beings die from COPD each year. Adults in midlife or older are maximum probably to enlarge COPD.

COPD consists of chronic bronchitis, emphysema and bronchial allergies. Many people have COPD and don't recognize it.

The individual with COPD reviews coughing, shortness of breath, tightness of the chest, and wheezing. Coughing generates a immoderate volume of mucus.

COPD blocks airflow within the lungs. The sickness is present day, because it makes respiration more hard.

Breathing is tougher because airways and air sacs come to be masses a great deal much less elastic. Walls among air sacs are broken or destroyed, and are inflamed and thickened. They end up clogged with the beneficial aid of the large amounts of mucus.

The breathing troubles contribute to a faded functionality to do day by day topics maximum folks do without questioning. It can be tough for the affected individual to put together dinner, stroll, and address themselves and their households.

Stem cell treatment has anti-inflammatory houses which could enhance lung feature. Stem cellular treatment protects damaged lung tissue and helps it to regenerate. Stem cells may also suppress infection in broken lungs, and can boom boom issue manufacturing.

In rats, studies has demonstrated development in pulmonary emphesema. Alveoli and blood vessels inside the lungs additionally professional quicker regeneration after lung amount reduction surgical remedy has been achieved on rats.

An article on PubMed said on a look at wherein sixty patients showed top notch effects. No excessive terrible outcomes happened after infusions of stem cells, while the sufferers were given 4 month-to-month infusions. The sufferers have been evaluated for 2 years after the primary infusion. None had worsening of their COPD.

The study concluded that stem cell treatment seems to be steady for mild-to-excessive COPD.

Mouse model studies have indicated that lung tissue harm from infection may be decreased via stem cells. Mouse studies utilizing character stem cells were visible to preserve giant improvement to COPD and special respiratory ailments

Up till the advent of stem mobile remedy, a affected person's only options have been treatments that tried to restrict harm and relieve signs and symptoms as a bargain as feasible. But now, as man or woman stem mobile treatment manifests immune and anti inflammatory benefits, the possibilities for human beings with COPD are extensively brighter.

Diabetes and Stem Cell Therapy

People with diabetes have levels of blood sugar (glucose) that are too immoderate. If glucose degrees live too excessive for too lengthy, fitness problems may have an effect on their eyes, kidneys and nerves. It can bring about coronary coronary heart sickness, stroke and, in intense cases, amputation of limbs.

Type 1 diabetes is also referred to as juvenile-onset diabetes. The immune system assaults its personal cells. The islet cells of the pancreas, even as healthful, manufacture insulin. However these cells are destroyed with the useful resource of the autoimmune destruction.

When insulin isn't pleasing its feature, glucose builds up in the blood.

Type 1 diabetes is handled via insulin injections often via the day. Sufferers need to check their blood glucose ranges 3 or four instances every day.

This is a existence sentence. When monitoring isn't carried out often and usually the individual with type 1 diabetes is at threat for coronary heart sickness, retinopathy, among different headaches.

Type 2 diabetes is also referred to as person-onset diabetes. Older folks which are obese and inactive, who have a family data of diabetes, are maximum susceptible to kind 2 diabetes.

In the ones instances, insulin is not used successfully via the body. This insulin resistance reasons accumulation of glucose in the bloodstream.

Type 2 diabetes typically can be controlled through weight loss program, exercising and taking treatment orally. The disease can improvement even as now not monitored because it must be. Insulin therapy can turn out to be critical.

Historically, remedy of diabetes has been restrained to life-style techniques in conjunction with bodily exercise, protection of a wholesome weight, and cautious food alternatives. Monitoring glucose tiers is critical for all diabetics. Some should additionally take treatment.

Stem cell therapy might be the final solution to treating and curing diabetes. The cells might multiply and differentiate in vivo to create the types of cells desired.

In April, 2014, it changed into delivered that the number one sickness-particular line of embryonic stem cells from a affected character's DNA were created. Such stem cells embody the affected character's private genome. Stem cell treatment can also need to replace any cells which can be broken.

It can be extra effective to use stem or precursor cell kinds which have the capability to be cultured to produce each form of islet cluster cells. This must create a proliferation of cells that would modify the amount of insulin to be released according to provide glucose tiers within the bloodstream.

Unfortunately, type 1 diabetes can cause an autoimmune response thru manner of the immune device to the beta cells. Induced pluripotent stem cells or iPS cells might be formed from person stem cells, after which converted into one of a kind forms of cells collectively with beta cells.

Beta cells, which is probably islet cells making insulin, or some one-of-a-type type of pancreatic islet cells, also can fill the bill. Isolated beta cells do no longer reply as effectively to glucose awareness modifications as do intact islet clusters of all islet cell sorts.

Beta cell response to glucose can modify itself. Very immoderate glucose levels need to necessitate the release of greater insulin at a quicker price. More slight glucose tiers couldn't want a lot insulin so rapid.

Stem cells determined within the biliary tree (which includes drainage ducts connecting the liver and pancreas to the gut) are pancreatic precursor cells within the manner of becoming pancreatic cells.

Biliary cells change into islets that create c-peptide and insulin in response to glucose within the blood. In mouse fashions, the ones cells have confirmed advanced manage of glucose. These cells might be the only vicinity from which to reap beta cells.

The scenario is a complicated one, however the rewards of being able to use stem cells to treatment diabetes can be outstanding. No greater headaches due to extended blood sugar. No greater finger-pricking, no more having to maintain such near track of what changed into eaten and at the same time as.

For the diabetic, stem cell therapy represents a new hire on lifestyles.

Erectile Dysfunction and Stem Cell Therapy

Treatments for erectile disease have left a few thing to be preferred. For this reason, the purpose of stem cell remedy for ED is pretty appealing. It might also want to heal the harm, as inside the instances of post-prostatectomy erectile sickness, or ED because of diabetes, or simply growing older.

Oral medicinal drugs increase blood drift to reason an erection. They also can have factor outcomes collectively with nasal congestion, flushing, gastric dissatisfied, headache and imaginative and prescient issues.

Not all guys can take the medication. For guys who can, the medication don't commonly paintings, and on occasion once they do artwork, they don't art work satisfactorily.

Self-injection of drugs into the penis can result in an erection that lasts for an hour. As daunting because the possibility of self-injecting can be, the pain is minimal. However, bleeding and fibrous tissue formation at the net website online of injection are possible facet consequences.

Penile suppositories can reason an erection internal 10 mins that would very last for half of of an hour to an hour. But ache, urethral bleeding and fibrous tissue formation may be component consequences.

Testosterone alternative remedy may also additionally help guys who have low testosterone degrees.

Penis pumps can reason an erection. They also can reason bruising. Climax may be much less than fulfilling. Those presented in intercourse classified ads need to now not be trusted. See your medical doctor.

Surgical penile implants are inflatable or semi-inflexible rods thru which a person can adjust his erection. This is a unstable alternative which can also cause contamination.

It's easy to look why stem cellular remedy for ED is probably superior treatment for affected guys.

Men who've had prostate maximum cancers can be helped through stem cellular treatment. Prostate cancer is the second maximum not unusual reason of loss of existence for men inside the United States.

It is also the most not unusual maximum cancers amongst American guys.

Prostate maximum cancers is turning into a extra commonplace terminal infection in the relaxation of the area. Prostatectomy, that is done to deal with prostate maximum cancers, is the number one reason for harm to the cavernous nerve.

This situation is a common after-impact of harm to the cavernous nerve that promotes healthy feature of erectile tissue. In truth, 20 to 70 percent guys who've had prostatectomy will have ED later on.

MSC transplantation seems looking in advance to men with positioned up-prostatectomy erectile disorder.

Adult stem cells are getting used to regenerate nerves which have been injured. Research on rats indicated that internal strain at the cavernous tissue become more applicable considerably.

Injections of ADSCs introduced improvement to shut to-everyday in research for 50 percent of the rats.

Erectile dysfunction is often a end end result of developing older and diabetes. Research for remedy of fellows who are diabetic, or whose age is a issue, has mounted that stem cells stayed possible for 21 days or longer after corporal injection.

Endothelial-derived NO/cyclic guanosine monophosphate signaling improved.

Research concerning animal studies indicated that ADSCs which had been grown in culture had effective effect as cavernous nerve damage treatment. Rats who have been injected with person stem cells had prolonged-time period recovery of nerve device functionality.

Frailty Syndrome and Stem Cell Therapy

Aging is a natural technique that all of us experience, and the older we get the greater recommended its consequences on our lives. Stem cells become less green at appearing their jobs in preserving our fitness.

It's not that we are losing stem cells, it's actually that they may be tons much less probably to upward thrust to the occasion than they was. Frailty is one of the consequences of this process.

Frailty isn't handiest a description, in which geriatrics is worried. Frailty syndrome is a geriatric assessment that relates to fine tendencies which includes excessive fatigue, exhaustion, generalized susceptible factor and unexplained loss of weight.

These tendencies can be measured by using the usage of assessing topics much like the affected character's strength and hobby levels, grip power, speed of gait and weight loss.

Frailty syndrome is due to a discounted ability to address physical stressors. Energy, protein reserves and electricity are less than they became. Multisystem decline has an regular impact at the affected character.

Adults over sixty five years of age have a better risk than regular for things like excessive toxicity or lack of lifestyles from chemotherapy for cancers. This is likewise actual for those who have frailty syndrome, regardless of the age.

It especially holds proper for sufferers with frailty syndrome who're confronted with transplantation, threat for infections, next health center stays, complications and medicinal capsules used. The prospect of stem mobile remedy may additionally deliver with it a long way decrease risks.

Arthritis, coronary heart failure, pulmonary infection, pores and pores and skin troubles, stroke and vascular disease are a number of the conditions that would rise up due to frailty.

Chronic inflammation is a deliver of revolutionary unwell health. Stem cellular remedy can be used to decrease this infection.

Regenerative medicinal drug like stem cellular remedy can restore damaged tissue, enhance the immune device, and heal chronic fitness problems.

Adult stem cells that are harvested and manipulated ex vivo can cope with frailty syndrome and different age-related fitness issues. Stem cellular remedy may additionally want to make cells extra inexperienced yet again.

Stem cellular treatment can make the distinction for people with frailty syndrome in very concrete, everyday techniques. For a few, it could suggest that they could walk without trouble over again, or that they may be capable of prepare their very personal food, and cope with themselves like they used to.

Chapter 5: Liver Failure And Stem Cell Therapy

The liver, situated on the right issue of the belly, plays a essential function in digestion and cleaning of the body. If the liver isn't functioning nicely, many important fitness conditions can crop up.

Problems can arise from starting with genetic abnormalities, and also from liver harm that happens due to factors like alcohol and viruses. Liver damage can cause cirrhosis, or scarring, which can result in the lifestyles-threatening condition liver failure.

Liver disease can be as a result of viruses and parasites. Inflammation can quit result, and the liver ceases to function efficiently.

Viruses spread thru blood, infected meals or water, semen, or contact with someone with contamination. Hepatitis is the most common form of liver infection. This magnificence includes hepatitis A, B and C, similarly to abnormalities to the immune device.

In autoimmune liver illnesses, the immune system assaults the frame, and on occasion this influences the liver. Autoimmune hepatitis, number one biliary cirrhosis and number one sclerosing cholangitis are amongst this beauty of disorder, similarly to genetics.

Liver illness is normally because of alcohol abuse and nonalcoholic fatty liver disorder (NAFLD) that is as a consequence of the buildup of fats in the liver.

Liver disorder threat will increase because of various factors. Shared needles, frame piercings and tattoos are a few chance factors. Diabetes and weight problems, further to immoderate triglyceride levels in the blood could make threat greater.

Unprotected intercourse, exposure to some distinctive's blood and physical fluids, or to 3 pollutants and chemical substances, increase chance. Any blood transfusions received earlier than 1992 growth the threat.

A patient with liver disease can also moreover moreover have yellowish eyes and pores and skin from jaundice. Their skin may additionally itch and their ankles and legs may additionally moreover swell.

They can also enjoy exhaustion, nausea, and vomiting. They also can lack an urge for meals. Their stomach can be sore and swollen, urine can be a dark coloration, and stool may be bloody, mild or the coloration of tar. They can also moreover boom bruising resultseasily.

In some instances, changes in way of lifestyles, like abstaining from alcohol or losing weight, can also need to make a difference for someone with liver disorder. In more immoderate instances, remedy or surgical remedy may be required.

Diseases that might reason liver failure have first-class the choice of liver transplantation. But liver transplantation is best a partial answer when you consider that out of 18,000 human beings searching ahead to a liver, best approximately 5,000 will obtain one.

Mouse version research has confirmed that mice with acute liver damage who had stem mobile treatment with realistic hepatocytes made from human stem cells, survived, as human liver cells had regular feature.

The future possibilities for stem cellular remedy getting used to regenerate a liver is interesting.

Stem mobile treatment using embryonic or grownup cells might also need to probably treat human beings with congenital liver issues, and will hold those tormented by acute liver failure alive, trying to find time for the patients to get better. Viral hepatitis may also be treatable through using way of stem cells.

Multiple Sclerosis and Stem Cell Therapy

In the past, a analysis of a couple of sclerosis have been one that had no get away direction. Some instances of MS are slight, a few are excessive.

Multiple sclerosis is a disorder of the immune gadget. The myelin sheath that is meant to shield nerves is beneath attack by means of using the character's immune device.

Some signs and symptoms also can moreover seem and disappear often without any apparent purpose, yet extraordinary signs and symptoms and signs and symptoms can also live constant.

The man or woman with MS has no manner to understand whether or not or not their signs and symptoms turns into worse as time is going on, whether or not or now not any enhancements will live or be out of place.

Stem mobile treatment offers desire for repairing and regenerating harm to the huge frightened tool from more than one sclerosis. Other remedies at exquisite ought to in all likelihood reduce irritation.

The National MS Society endorses research and development with person stem mobile treatment. Stem cell remedy also can slow down the development of MS, and repair worried device harm. As but there aren't any permitted MS remedies the use of stem cells.

One technique being studied includes the transplantation of autologous hematopoietic (blood cell-producing) stem cellular transplantation (HSCT) in an try to restart, or reboot, the immune system.

Success on this project, it is was hoping, will open the way to protecting brains and spinal cords from in addition damage from MS.

Ultimately the aim is to replace all the frame's immune cells with new immune cells thru stem cells, with a view to no longer assault the man or woman's mind tissue, consisting of myelin.

Another technique being tried isolates and multiplies mesenchymal stem cells, then puts the cells again into the individual's body in preference to changing immune cells.

Yet every different method is supposed to healing damage to the worried tool, the use of stem cells to update myelin-producing cells further to damaged nerve cells. It is theorized that this can work so long as precipitated pluripotent stem cells (iPS) are not rejected with the resource of the immune tool.

Autologous mesenchymal stem cells are being studied. Research is being carried out to evaluate whether or now not it's far steady and viable to carry out intrathecal and intravenous management of those stem cells, which is probably additionally referred to as mesenchymal stromal cells.

MSC transplantation in people who've MS can be solid and powerful in changing and regulating the immune tool. Research appears to indicate that MSCs are a feasible device for reducing contamination and restore through integration of tissue, and neural cellular differentiation.

MSCs appear to promote advantages in each continual and acute animal models of MS. Research is being completed to decide how secure and powerful the usage of those cells might be in shielding the apprehensive system for secondary revolutionary MS.

These cells display anti-proliferative residences (inhibiting cell growth). They prevent neuron lack of lifestyles and reduce infection.

Stem mobile remedy is in its infancy, however the destiny seems vibrant for those who are currently managing more than one sclerosis.

Parkinson's Disease and Stem Cell Therapy

More than 1 million Americans be bothered through manner of Parkinson's sickness. This is a neurodegenerative illness which includes an ongoing lack of dopaminergic (DA) neurons, a contemporary accumulation of Lewy body inclusions, and dysfunction in motion.

PD reasons trembling fingers, legs, face and jaws, in addition to inflexible or stiff trunk and limbs. Slow actions, and risky balance, coordination and posture are also signs and symptoms of PD.

Other symptoms and signs that could stand up are emotional changes like depression, incontinence or constipation, and problems with the pores and pores and pores and skin. Difficulties with chewing, swallowing, speakme and napping may moreover arise.

PD is identified in step with neuorological exam and clinical information because no blood exams or laboratory reviews can understand it. This makes a right prognosis hard.

Parkinson's sickness is classed as each continual and contemporary: chronic in that it's miles a protracted-term situation, innovative in that signs end up worse through the years.

Walking and speakme end up extra difficult. Simple moves and actions emerge as harder to do. Tests may be run to rule out other situations.

Parkinson's disease usually appears after 60 years of age.

The severity of signs and symptoms and signs that can be professional varies from affected person to affected character. Some can also additionally never have more than moderate symptoms and signs, at the same time as others will bypass at once to managing excessive lifestyles-converting deterioration. Will the individual with moderate signs and symptoms and symptoms moreover pass right now to grappling with more immoderate disruption? No you may say.

Traditionally remedies for Parkinson's disorder have only been able to mood a number of its symptoms. Thus an prolonged manner, remedy has no longer been to be had that could prevent the development of the sickness or contrary it. Healing, lamentably, has simply no longer been viable. Yet.

Preclinical stem mobile treatment studies are aimed in the direction of converting malfunctioning neurons in the destiny. It's viable that such remedy alternatives may additionally moreover protect DA neurons and facilitate their regeneration, specially thru using mesenchymal stem cells.

Adult human endometrial-derived stem cells are a with no trouble obtainable kind of mesenchymal stem-like cells. And HEDSCs had been efficaciously transplanted and have been used to generate dopaminergic cells, the dearth of which has added about such disastrous signs and symptoms and signs and symptoms for humans with Parkinson's disease.

HEDSCs were located in an immunocompetent PD mouse version to be likely appropriate property of allogenic stem cells. Recouping concentrations of dopamine has been promising.

Cellular fashions of diverse types are being superior with iPS cell-derived DA neurons using animal fashions. Induced pluripotent stem (iPS) cells might also make feasible treatment customized to the person affected character.

Reprogramming generation are below improvement which might be non-DNA-mediated, non-integration and non-viral. Developing higher non-viral/non-biased strategies of reprogramming, and producing homogenous midbrain DA neurons is a goal.

Ultimately it's far hoped that PD will in the future be successfully handled with a aggregate of cell alternative, gene treatment and suitable drug remedies, in order that PD's development may be stopped in its tracks for suitable.

Pulmonary Fibrosis and Stem Cell Therapy

Pulmonary fibrosis is a type of interstitial lung infection. It generally impacts the insterstitium (tissue and place surrounding the lungs' air sacs). It doesn't right away effect the blood vessels or the airways. But due to the fact the parenchyma (beneficial elements of the organ) is regularly however inexorably modified with scar tissue, life gets extra hard, and wish for a better destiny will become hard to come with the aid of.

Deep lung tissue becomes scarred, and will become stiff and thick. "Fibrosis" manner "scarring," referring proper here to the lungs. Breathing is greater tough to do. Blood won't be receiving appropriate enough oxygen. Scarring will boom over the years.

Pulmonary fibrosis may also get up because of some connective tissue illnesses. It can arise due to environmental pollutants. Interstitial lung illnesses are pretty pretty a range of sicknesses that scar or inflame lungs. They can result in PF.

A character with pulmonary fibrosis may moreover additionally experience a dry hacking cough, and shortness of breath. Other viable signs and symptoms are aching joints and muscle groups, clubbing (widening and rounding of toe and finger recommendations), fatigue, and unexplained weight reduction.

Diagnosis of PF is made with interest of clinical information, biopsy, imaging exams and tests for lung feature. There isn't any appeared treatment for pulmonary fibrosis, even though remedy can be determined from a few symptoms. It is hard to discover tablets as a manner to paintings in every case of PF, but, in factor because of aspect results that may rise up.

Stem cell remedy want to change all this, when you maintain in thoughts that stem cells can redecorate themselves into any form of mobile within the body.

MSCs is probably promising for treatment of PF as they aim damage internet websites and reduce contamination, and assist to restore epithelial tissue. The MSCs secrete multifunctional antifibrotic (causing regression of fibrosis) humoral factors (transported with the aid of way of the circulatory gadget), and can proper the balance among epithelial and mesenchymal cells via Stanniocalcin-1 being secreted in a bleomycin-prompted pulmonary fibrosis version.

MSCs have finished properly in clinical trials. In a few preclinical research in animal fashions consequences have recommended that stem cells can also address pulmonary fibrosis correctly.

Adipose stem cells have been applied in animal models of emphysema. The ASCs were discovered in the parenchyma and lung airlines. They extra suitable inflammatory infiltration and cell loss of life. More research is wanted, but the opportunities are encouraging.

Chapter 6: Renal Failure And Stem Cell Therapy

When the kidneys are not able to put off waste, and might't nicely preserve fluids and electrolytes in stability, kidney failure can be the end result.

Bloody stools, a metallic flavor in the mouth and smooth bruising are a number of the symptoms that may accompany kidney failure. Decreases in urge for food, energy and sensation in the extremities, collectively with swelling in the ankles, toes and legs, also can stand up. Other signs and signs are problems with respiratory, bleeding, further to gradual motion and tremors.

At the existing time, treatment for kidney failure is restrained, and regularly involves time in the hospital. Restrictions commonly ought to are available to play. The amount of drinks is constrained. Many components are removed from the safe-to-devour list.

Infection prevention or remedy might also require antibiotics. Fluid buildup can also necessitate diuretic use. Blood potassium tiers may additionally moreover need to be regulated through using IVs. Dialysis can be needed, every for a quick time, or completely.

These are all true reasons to be focused on research into stem mobile remedy for renal failure.

Embryonic and grownup pores and skin stem cells have been transformed into kidney tubular cells with the resource of researchers from Brigham and Women's Hospital and the Harvard Stem Cell Institute in Boston. Stem cells expressed key markers of early kidney-forming cells, the intermediate mesoderm. These cells then expressed a marker that's essential for kidney differentiation, Six2.

Results have been confirmed whilst researchers in Australia, Japan and the united states have been also capable of transform stem cells into Six2 powerful cells.

Researchers are hopeful that around the following nook may be enhancements in treating kidney illnesses which consist of polycystic kidney illness, in addition to drug toxicity regarding kidneys.

Stem cellular remedy is expected that could some day broaden without a doubt useful kidneys, which could probably recommend advanced remedies for continual kidney sickness.

Research has been completed in the quest for logo spanking new remedy plans for continual kidney ailment. Rat model studies have had success with adult bone marrow stem cells.

A take a look at at St. Michael's Hospital indicated that enriched stem cells can beautify rats' kidney feature. Hormones were secreted through enriched bone marrow stem cells. These hormones had been injected into the rats, and had the equal effect as stem cells might have.

The gain of those findings is that the advantages of stem mobile treatment can be received whilst no longer having to inject them.

End-diploma renal failure and continual kidney sickness affects increasingly human beings because of population growing antique. Such treatment plans might be critical in worrying for such plenty of.

More than 20 million Americans have a few stage of continual kidney disease. Restrictions in every day lifestyles are their fact, alongside aspect medicinal drugs, ache, and in more immoderate instances, dialysis.

In the most extreme instances, kidney transplantation can be the most effective solution. Presently, many are watching for kidney transplants but the fashion of donors doesn't preserve up with the range of those in need. And a transplant can occasionally be rejected via the immune system of its recipient. Immunosuppressive drugs will need to be taken for the duration of the relaxation of the recipient's existence.

Stem mobile remedies could be the open door to regenerative treatment plans that could heal and replace broken organs and tissue, making transplantation useless.

Rheumatoid Arthritis and Stem Cell Therapy

Rheumatoid arthritis takes place on the equal time because the immune device assaults the joints. Inflammation outcomes in thickening of the synovium (tissue lining the interior of joints), contributing to swelling and painful joints.

Inflammation can damage cartilage and bones, if not treated correctly. Joint damage that occurs is irreversible.

RA commonly assaults arms, elbows, wrists, toes, knees and ankles. Usually attacks are symmetrical, so if one joint on one factor is affected, the corresponding joint on the opposite issue of the body is often moreover affected.

RA is a systemic illness due to the fact it is able to have an impact on body structures like the breathing tool or the cardiovascular system.

According to The Arthritis Foundation, about 1.Five million Americans have RA.

Treatment of rheumatoid arthritis is excellent if it's far early and aggressive. The spread of infection should be halted to shield inside the direction of ongoing damage. Disease interest (RA-generated contamination) need to be introduced into remission.

Historically, drug remedies had been relied upon to reduce signs.

Nonsteroidal anti-inflammatory tablets (NSAIDs) e.G., ibuprofen, ketoprofen and naproxen sodium, decrease ache and contamination. These are contraindicated for folks who can be susceptible to belly ulcers. In such instances, celecoxib (an NSAID known as a COX-2 inhibitor) can be prescribed.

Corticosteroid pills, which includes prednisone, prednisolone and methylprednisolone, are used sparingly to govern infection, due to excessive component effect dangers.

Disease-editing antirheumatic drugs (DMARDs) like methotrexate, hydroxychloroquine, sulfasalazine, leflunomide, cyclophosphamide and azathioprine, are implemented in an try and modify the course of RA.

Biologics, a subset of DMARDs, collectively with abatacept, adalimumab, anakinra, certolizumab pegol, etanercept, infliximab, golimumab and rituximab, block specific steps of the contamination machine. For a few with RA, biologics may be a hit in slowing, altering or maybe halting RA.

In the maximum severe times, surgery inside the shape of joint replacement has been an opportunity.

Fortunately, autologous hematopoietic stem mobile transplantation (HSCT) has been used for excessive RA.

Stem cells suppress a complex contamination mechanism. In rheumatoid arthritis, T cells attack the affected character's non-public tissues. Stem cells derived from fats can help lower this inflammatory mechanism. Some sufferers see blessings in the first to a few weeks with stem cellular therapy.

Adult mesenchymal stem cells suppress effector T cell and inflammatory responses. Immune issues also can in time be successfully handled with the usage of adult MSCs.

Human adipose-derived mesenchymal stem cells (hASCs) can be key regulators of immune tolerance. They may be capable of suppress T mobile and inflammatory responses. They may also provoke antigen-precise regulatory T mobile activation.

Most individuals in a look at using the Autoimmune Disease Databases of the European Group for Blood and Marrow Transplantation (EBMT) and the Autologous Blood and Marrow Transplant Registry (ABMTR) to become aware of RA patients had top results.

More than half of of the people completed an American College of Rheumatology (ACR) 50 or extra reaction in the term of three hundred and sixty five days.

Chapter 7: Stroke And Stem Cell Therapy

Stroke impacts arteries in the thoughts, and people that cause the thoughts. It occurs on the equal time as a blood vessel is blocked through a clot or rupture, rendering the vessel no longer capable of bring vitamins and oxygen to the thoughts. As the mind is starved for blood and for oxygen, thoughts cells start to die.

Stroke is the 5th most common purpose of death in the United States. It is the primary motive of disability for Americans.

The outcomes of stroke will depend upon in which it takes vicinity and what kind of impairment there may be to blood flow.

Stroke at the right facet will have an effect on the left component of the body and face inside the form of memory loss, paralysis or problems with imaginative and prescient.

Stroke at the left element of the mind affects the proper thing of the body, in strategies that embody speech and language problems, reminiscence loss and paralysis on the proper issue of face and body.

Stroke in the again of the mind may have an effect on imaginative and prescient.

A stroke that takes vicinity inside the thoughts stem can have an impact on both or each aspects of the frame and might immobilize a affected man or woman from the neck down, and may prevent or cast off their capability to talk.

A stroke takes location on the equal time as bleeding or a blood clot takes location, hindering blood go with the flow to the thoughts. Brain cells will die inside the event that they don't get sufficient oxygen and other nutrients quick.

The American Stroke Association opinions that 137,000 people die inside the United States from stroke each year. Many extra humans stay on stroke, but come to be disabled.

Stroke ought to have an impact on someone's capability to talk, assume, do not forget and go with the flow, relying on the location of the thoughts that's been broken. A individual may be left with vulnerable point or paralysis.

Some losses can be lessened or reversed with enough physical remedy. But it has not been viable to heal the harm to mind tissue. Research seems to indicate that stem cell remedy may be changing all that.

Research in 2014 indicated that when stem cells were injected into stroke sufferers' brains, none had any horrible results to stem mobile transplantation. All 18 had less paralysis or vulnerable component inside the first six months. Two patients stepped forward substantially, and almost at once.

By day after today, the two had been already beginning to talk and walk all over again. This grow to be splendid, because of the reality their surgical methods had taken region more than years once they had their strokes.

Animal version studies for stem cell remedy had been rendering promising effects. This new studies is the diverse first concerning human beings.

Participants, a while 22 to 75, had blood clots that resulted in stroke that left them with a few paralysis or weak factor, six months or more previous to their stem mobile injections.

MSCs have been derived from bone marrow and evolved in a laboratory. No immune-suppressing drugs were used. No rejection took place.

Animal model studies seem to indicate that the stem cells reason the broken mind tissue to carry out at their most capability degrees.

Research from the University of Toronto indicated that an injection of a hydrogel together with stem cells boosted mind healing after a stroke, in mice fashions. Blindness in mice moreover modified into in part reversed.

Research in which human stroke survivors had been given stem cellular treatment has produced effective results. A three hundred and sixty five days after neural progenitor stem cells had been injected into damaged areas in their brains, some sufferers are once more able to flow into muscle tissue that had formerly been weakened or immobilized.

MSC transplantation in animal stroke models has been seen to decrease duration of infarct, and sensible consequences had been favorably affected. An unblinded have a look at concerning human MSCs of 16 patients the various a long term of forty one and seventy 3 years antique tested every grey and white remember similarly to blended lesions.

The kind of people reduced to twelve due to various factors. All 12 contributors suffered from hemiparesis. Five sufferers had aphasia.

Autologous bone marrow-derived human MSCs have been given with the aid of intravenous with out high-quality horrible consequences. A moderate itch, a few localized pain, a moderate fever and nausea, and decreased urge for food had been visible in character patients. There have been no uncommon mobile growths, CNS tumours, neurological problems, venous thromboembolism, systemic malignancy or infection later on.

While researchers were cautious to emphasise that they could not definitively exclude the possibility of placebo impact or bias of any type influencing interpretation of look at outcomes, notwithstanding the truth that findings have been maximum encouraging concerning the benefits that might be accumulated from stem cell remedy for stroke patients.

Peripheral Vascular Disease (Vascular Insufficiency) and Stem Cell Therapy

Whether you name it vascular insufficiency, peripheral vascular ailment, peripheral artery illness, or peripheral arterial sickness, this circumstance is a distress to live with.

PAD is an indication of extensive fatty deposits within the arteries. In this form of case, your blood go together with the flow isn't always in reality being decreased on your legs, but can also be diminished on your thoughts and your coronary coronary coronary heart.

Plaque builds up in arteries to move, limbs and organs similar to the kidneys and the stomach. Arteries of legs and toes become narrower and harden. Blood drift decreases. Nerves and particular frame tissue may be damaged.

PAD can bring about cramping and ache at night time time similarly to numbness and tingling in ft and toes. Legs may be numb, prone, or cold with little or no pulse, because of the scenario. Hair at the legs and toes may additionally moreover broaden more slowly, or fall out.

Impotence, sores that don't heal, ischemic ulcer (open sores) on decrease legs, may additionally additionally moreover boom. Skin can deal with a darkish or blue coloring. Even the contact of sheets or garb can be painful.

PAD may be on account of atherosclerosis. It additionally may be due to infection of blood vessels, injuries to fingers or legs, or radiation.

Peripheral artery illness is a state of affairs that normally emerges in men over the age of fifty. It's located thru achiness, burning, fatigue and pain in the leg, calf, foot and thigh muscle tissues within the course of exercising. It is normally relieved thru the usage of resting for a couple of minutes. Over time, symptoms and signs come on greater fast and treatment more slowly.

Also at hazard for PAD are human beings over age 70, in addition to people over 50 with a facts of diabetes or smoking. People beneath age 50 can also get PAD within the occasion that they have diabetes and different PAD risk factors, like immoderate blood strain or obesity.

If you have got got PAD, you're at better hazard for infection, and might have greater problem fighting infection. In very intense times, gangrene or tissue demise can arise. Amputation can test.

PAD will increase danger for coronary coronary coronary heart illness, additionally known as coronary artery sickness, coronary heart attack, stroke and short ischemic assault or mini-stroke.

Treatment for PAD can slow the infection, and decrease the threat of complications growing.

The not unusual treatments are modifications in lifestyle, tablets and surgical operation. If you consume healthy, get enough exercising, and don't smoke, you may be capable of deal with PAD.

PAD can be dealt with with angioplasty and stents. But approximately 10 percentage of patients with PAD are not capable of have the ones remedies.

Revascularization is used to restore blood circulate to a place. It unblocks disrupted or obstructed blood vessels, or implants replacements thru surgery. When a fulfillment, this can hold a limb. Sadly, very regularly amputation and rehabilitation are the following step.

But stem cellular therapy may be the pleasant feasible solution. The aim of an infusion of stem cells will be to maintain blood drift shifting at a healthful fee.

Stem cells injected into their legs might also additionally assist human beings with PAD boom the form of blood vessels. These vessels need to take the region of damaged vessels with a buildup of plaque.

Animal fashions have proven that stem cells may also make it viable to construct higher blood vessels. This ought to make for a greater suit blood deliver for toes, and might reduce the probabilities of needing amputation. Stem cells can be directed to in which they will be most first-rate, and this could be carried out as generally as wished.

Chapter 8: Stem Cells: What They Could Do

You sure preferred you had the awesome recuperative powers of X-Men's Wolverine at one time or every different along with whilst you had a painful wound due to a carrying hobby or a car coincidence. While the recovery and regenerative functionality of Wolverine is a chunk of fiction with the aid of way of using the legendary superhero author Stan Lee, be aware that people also inherently own this option, however the truth that in a constrained revel in.

A man or woman continuously regenerates blood, skin and different frame tissues. However, humans cannot replace a lacking finger, an amputated leg or unique body additives on their non-public. But with the cutting-edge discoveries within the issue of drugs, the outcomes of these issues may be correctly addressed.

The stem cell therapy is really one of the many breakthroughs within the challenge of medicine. It is being broadly seemed as a treatment opportunity which offers a renewed want for the ill. Research on stem cells has truly attracted splendid hobby inside the past decade or so. The hopefuls take a look at those building blocks of existence as huge players within the remedy of persistent, degenerative and incurable illnesses, maximum especially osteopathic injuries.

They have numerous reasons to wish for from the scientific and clinical opportunities of the emerging technology as it moreover gives quicker remedy from the ache because of sports sports-associated and other osteopathic injuries. But, what actually are stem cells and why are researchers so beaten with the useful resource of what the ones unspecialized cells can do for the frailties of mankind?

Stem cells have the functionality to increase numerous mobile sorts within the human frame. These cells additionally function a form of inner tool which divides basically with none barriers to replenish broken or useless cells. A cell advanced through stem cellular department has the functionality to live as a stem cell or develop into each exceptional type which has extra specialised talents. Examples of those specialised cell sorts are crimson blood, mind and muscle cells. Stem cells moreover have number one defining abilities. These are:

•Self-renewal: Based on a few internal inter-cellular verbal exchange, a stem mobile divides at the same time as it receives the sign to achieve this and regenerates or self-renews to carry out particular duties together with tissue restore.

•Reform capability: Once transplanted into one a part of the body, a stem cellular simply reforms the suitable tissue wherein it emerge as routinely or deliberately transplanted to perform its vaunted motion.

The above features make stem cells the maximum promising medical leap forward to date. Bone marrow studies in the 1950s paved the way for in addition investigations on those fantastic constructing blocks of existence. In the early 1960s, scientists laid down the muse of stem cellular technological statistics.

There are most essential education of stem cells:

•Embryonic and fetal stem cells

•Adult stem cells

Both training are similarly vital. Further research and research added about the invention that every of those cells paintings in conquering ailments and alleviating positive conditions. Try googling key phrases associated with stem cells and your search will turn up a results page that shows what fields of medication have drastically benefited and are continuously locating succor in the therapy. A lot of medical practitioners are clearly the use of stem cells to alleviate the signs or address patients with the following situations:

•Alzheimer's illness

•Blood issues

•Cancer

•Celiac disease

•Deafness

•Diabetes

•Heart illness

•Huntington's disease

•Immunology and autoimmune illnesses

•Infertility and reproductive problems

•Metabolic illnesses

•Multiple sclerosis

•Muscle harm

- Neurological troubles

- Osteoporosis

- Parkinson's disorder

- Retinal degeneration

- Spinal cord and one-of-a-kind osteopathic accidents

- Wound recuperation, and plenty of others.

What Stem Cells Can Do?

Through the technique of transplantation, stem cells can perform obligations that remedy diverse illnesses and specific health situations. The treatment particularly gives the subsequent blessings:

- Facilitates multiplication of the host stem cells and goad their feature so you can save you in addition damage or lack of tissues

- Fuses with host cells to useful resource in harm restore

Chapter 9: Autologous Stem Cells And Their Role In Healing And Recovery

The development of autologous stem cell treatment end up to start with completed using the bone marrow transplant way considering at that time, stem cells can best be sourced right now from the bone marrow. In the number one transplants performed, the medical doctor performs the bone marrow harvest, a gadget wherein stem cells are collected from the marrow of the donor and later infused into the affected character-recipient. This is accomplished in a clinic's on foot room with the donor beneath stylish anesthesia. The bone marrow is aspirated1 with a completely unique needle and the medical doctor filtered the stem cells harvested. Finally, the filtered harvest is infused into the bloodstream of the affected person similar to a blood transfusion.

Later on, technology became advanced for peripheral stem cell transplantation in which the medical medical doctor actions stem cells out of the donor's bone marrow into the peripheral blood. By peripheral blood is supposed blood cells which includes pink blood cells, white blood cells and blood platelets which can be circulated in the course of the body, and not blood cells inside the body's bone marrow, liver, lymphatic gadget and the spleen.

Doctors perform the transplantation via a technique referred to as apheresis2, it's miles facilitated by using using way of the use of substances known as granulocyte-colony stimulating element (G-CSF) or granulocyte macrophage-colony stimulating factor (GM-CSF). These substances are growth elements that are implemented to set off the manufacturing of cells.

The more moderen generation for stem mobile transplantation is plenty much less complicated on the donor because:

•Hospital admission isn't always required

•General anesthesia want not be used

•There is normally no pain after the technique

When a patient gets the transplant from his or her very very own wholesome stem cells, the transplant approach is known as autologous. There are distinct strategies wherein stem mobile transplantation is accomplished and the choice of gadget relies upon at the illness, the scenario of the patient and the deliver of donor. In an allogeneic transplant, the deliver of stem cells is a donor whose tissue kind is a close to match to the affected character's type which includes a brother, sister or at instances, someone else who isn't a relative but with a exquisite tissue type healthful. Syngeneic transplants are the least normally executed because of the reality the donor right here is an identical twin.

———————————

1 The device of aspiration entails getting rid of the cells thru placing a needle into the tissue and drawing the cells into a syringe.

2 Apheresis is the device of disposing of blood from a donor for processing with the reason of setting aside decided on factors of the blood and returning the unharvested factor lower back to the donor.

Autologous stem cell transplantation, it in reality is the modality defined in this ebook, is logically the superb technique of preference if there aren't any hindering conditions to be able to save you its use. Patients who achieve their very non-public cells all over again are better off than the ones receiving the transplants from a donor. An autologous transplant does now not positioned the recipient vulnerable to the graft-instead of-host disease (GVHD). The manner additionally assures the patient that she or he could be capable of not get a contemporary contamination or disease from the donor. The feature of autologous stem cellular transplants in healing and recuperation is initially restricted at present time as an experimental treatment desire for the subsequent ailments:

•Autoimmune issues together with Crohn's sickness, lupus erythematosus, a couple of sclerosis, rheumatoid arthritis, and so on.

•Certain leukemia

•Dysgerminomas

•Lymphoma

•Multiple myeloma

•Osteopathic injuries

•Solid tumors

•Some cancers which incorporates neuroblastoma, testicular most cancers, and some types of pediatric maximum cancers

•Systemic sclerosis

•Teratomas

Autologous stem cellular transplant does not bring about GVHD. This modality of transplantation has now not stated times in which the transplanted stem cells treat the frame of a affected person as a distant places matter and attacks the body. The pinnacle judgment right here is that for the cause that stem cells are from the equal person, they do no longer flip in the direction of their very non-public kind. This is an crucial function that the autologous technique performs inside the remedy of the above-mentioned diseases.

Chapter 10: Autologous Stem Cells And Osteopathy

The possibilities of recuperation and healing via the growing technology, autologous stem cell remedy, seem like constant. A lot of researchers prove thru their research that this treatment will rapid emerge as the idea for curing numerous ailments like Parkinson's disorder, coronary coronary coronary heart failure, coronary coronary heart disorder, cerebral palsy, diabetes and a collection of other vital ailments and osteopathic accidents.

According to investigate, stem cells also are useful for those who plan to reveal show new drugs and pollutants and understand begin defects without the need to problem human volunteers to volatile tablets and pollutants. With an extended roster of health situations that might advantage from this terrific stride in scientific technological information, humans can't assist asking how and why the arena of osteopathy is leaping into the bandwagon of autologous stem cell treatment hopefuls.

In the usa, the Food and Drug Administration permits using bone marrow stem cellular transplantation system as a shape of remedy. This is a great trouble for patients who would love to strive out this treatment. The approval obtained thru this treatment from numerous authorities bodies and clinical establishments inside the US show that it is steady.

For individuals who are stricken by osteopathic accidents, autologous stem cellular remedy also can be a massive assist. The region of osteopathy consists of the recuperation of smooth tissues for the proper manipulation of bones and muscle groups which might be laid low with an damage. The magic word inside the recuperative courting among an osteopathic damage and the stem cell is 'tissue' - gentle tissue.

Soft tissues are those who be a part of bodily organs and unique systems or people who surround such organs and systems as blood vessels, fascia, fats, fibrous tissue, ligaments, muscular tissues, nerves, pores and pores and skin, synovial membranes, tendons, and so on. Soft tissues that comprise of cells and stem cells can restore damaged cells and tissues. Once gentle tissue repair and damage manage are completed, osteopaths can do their difficulty. Among the subjects that they are able to do with the beneficial resource of autologous stem cellular treatment are the following:

•Manipulate the injured vicinity(s)

•Promote mobility now not in reality in phrases of the injured component but the complete frame as a whole gadget

•Restore the herbal balance within the frame that turn out to be compromised due to the damage.

Chapter 11: Autologous Stem Cell Therapy And Its Ability To Heal Joint And Other Osteopathic Injuries

Autologous stem cellular transplantation entails the collection, harvesting or isolation and processing of the aim stem cells, and the re-injection to the identical man or woman from whom the cells were taken for recuperation purposes. The processed stem cells may also be saved for later use with the aid of the equal person.

Autologous stem mobile remedy normally includes transplanting the stem cells with the help of the PBSCT (peripheral blood stem cell transplantation) technique. This machine takes the stem cells out of your blood. The entire method additionally makes use of G-CSF protein, a boom detail, that is beneficial in stimulating new stem mobile boom. This allows the cells to spill into the blood. Note that G-CSF is actually produced with the beneficial aid of your frame.

The majority of times that encompass autologous stem cellular transplantation for adults employ stem cells taken from their blood. In the case of a child, the professionals and the child's circle of relatives will decide whether or not or now not to get and use the stem cells from the blood or from the bone marrow based totally totally on the child's length.

Healing Joint and Other Osteopathic Injuries thru Autologous Stem Cell Therapy

The number one effect of autologous stem mobile remedy is the right away regeneration of damaged bones, cartilage, ligaments, tendons and joints. This manner that the remedy is beneficial in curing numerous varieties of joint and osteopathic injuries. Stem cells can be predicted to offer advantageous advantages to the ones laid low with joint and osteopathic accidents. This is made feasible with the aid of stimulating and supporting their blood vessels and reparative tissues.

One of the most thrilling packages of autologous stem mobile remedy is the restoration of bone cysts and mild tissue joint accidents. The therapy moreover contributes in dealing with lameness generally related to osteoarthritis. The transplant is usually finished along aspect arthroscopic surgery, especially if there are horrible and unwanted results from the surgical procedure.

The capability of stem cells to remedy joint and osteopathic injuries can be attributed to their anti-inflammatory houses. The cells also are able to embedding inner your joints and supporting self-renewal. This regularly outcomes in gambling its terrific results for a extended period.

The present era moreover permits clinical practitioners to withdraw stem cells from your blood or bone marrow, method them in the laboratory and re-inject the cells into the injured joints and tissues indoors your frame. These techniques are commonly done with the assist of advanced imaging. With the assist of MSK ultrasound, Fluoroscopy and other advanced imaging techniques, the cells may be brought into the precise body areas that want hobby.

One gain of doing this is enhancing the natural restore manner of your injured and degenerated tendons, arthritic joints and ligaments. The terrific detail about the autologous stem cellular transplantation manner is that it is able to be finished inside simply someday. This is one of the excellent alternatives for sufferers who require surgery or joint opportunity.

Expect to head again for your ordinary sports activities after a brief rehabilitation length. This short rehabilitation length is usually had to repair your energy and mobility.

Chapter 12: Is Stem Cell Therapy In Osteopathic Injuries Safe?

At the existing time, stem mobile remedy from bone marrow transplant is authorized with the aid of the use of the FDA. This is the extraordinary guarantee of protection all people gets. The FDA has authorised stem cellular transplant techniques from autologous bone marrow or blood stem cells. Most of the transplants have been finished on sufferers afflicted with most cancers and severa disorders that affect the blood or the immune system.

With the consistent stem cellular-based totally definitely treatment it truely is FDA-authorized, plenty of athletes are given 2d possibilities to reclaim their supremacy inside the sports they excel in and masses of greater are blessed with the desire of getting returned into the movement for the fun of it. Stem cell remedy can be successfully utilized in treating sure situations which have failed to respond to extra conservative remedies.

The critical reason of the remedy inside the field of osteopathy is to inject adult stem cells into damaged body tissues to relieve pain, repair their everyday capabilities and sell restoration. With the capacity of the stem cells to obviously become healthful and new tissues, athletic and osteopathic injuries will virtually be repaired.

Just ensure to look for an expert and professional clinical clinical doctor who will perform the device of setting the gathered cells out of your frame in awesome areas wherein these are wanted the maximum, specifically the injured areas. The correct trouble approximately autologous stem cellular transplantation is that the cells are taken out of your very own frame, so the risk of tormented by component effects may be very minimal. In the sector of osteopathy, the application of individual stem cells can considerably assist parents which might be stricken by:

•Osteoarthritis affecting the joints

•Partial however persistent tendon tears – Examples of those are plantar fasciitis, patellar tendon tears, quadriceps and tennis elbow

•Severe partial rotator cuff tears

•Cartilage tears affecting the knee

•Partial muscle tears

To make certain your protection even as making plans to undergo stem mobile treatment, it's miles critical to analyze or ask a practitioner approximately the massive sort of durations which you want to get the amazing restoration consequences. In maximum instances, sufferers of osteopathic injuries respond properly to the treatment with handiest a unmarried consultation. For more excessive cases, to three remedies can be required.

Most patients who go through the machine also be conscious symptoms and signs of improvement inside four to 8 weeks when they have completed the required treatment protocols and large fashion of sessions. Some of the superb outcomes that they skilled after this era are less regular pain, faster restoration from painful accidents, and capability to carry out extra sports previous to experiencing the pain due to any harm.

Dealing with an incredible stem cellular treatment expert also can increase the favorable effects of the remedy because of the reality this expert can customise the remedy to fulfill your specific needs and scenario. This will similarly make sure your protection.

Success Rate

The number one goal of autologous stem cell remedy is to reduce the quantity of ache suffered with the resource of osteopathic damage patients. This will give up end result to improved feature in areas of the frame stricken by the damage. Based on studies, the fulfillment charge of the remedy is round 80 to ninety percentage for osteopathic harm and arthritic joint patients. This percentage is already excessive for a newly advanced treatment modality.

Some sufferers of the cited accidents even enjoy entire pain treatment after virtually one treatment. For ligament and tendon accidents, sufferers can expect eternal outcomes. For individuals who suffer from joint arthritis, the amount of training required or the duration of remedy can be dependent upon the severity of their case.

Mild arthritis and osteopathic accidents are curable within certainly one remedy. More continual instances of arthritis and osteopathic injuries, as a substitute, may additionally additionally require more than one remedy. A repeat remedy can be desired, generally in a single to a few years.

Chapter 13: The Autologous Stem Cell Solution To Osteopathic Injuries

How is the arena of osteopathy keeping out with conventional treatments vs. Stem mobile transplants? The results are thus far encouraging. While a single bankruptcy can not communicate the whole lot there may be to comprehend about the benefits of the stem cell choice to patients with osteopathic accidents, some examples can get the message for the duration of.

Sciatica

You wouldn't want to enjoy the pain of sciatica. The excessive ache in the decrease extremity is unbearable. This is typically traced to the infection inside the root of the sciatic nerve. Compression of the nerve root can also bring about sciatica pain. This state of affairs may be attributed to an overgrown issue joint, a herniated disc inside the lower lower back, protrusion of the disc or thickened ligaments.

The conventional treatment for sciatica includes epidural injections comprising of steroids and anesthetics introduced through guided X-ray, surgical procedure or bloodless laser treatment. The injections can numb the ache, but the motive of the pain is unmanaged and there can be typically a heightened chance of cellular lack of life. Surgery is pretty bulky due to the hospital confinement referring to the approach and as plenty as half a 12 months of rehabilitation. The non-invasive cold laser remedy does now not promise full consolation from sciatica signs and symptoms after one or perhaps 10 treatments.

The use of stem cellular transplants injected into the epidural location, as an alternative, can boom blood flow. This will in the end prevent the degeneration of the disease disc and contrary the harm. Stem mobile remedy does not cause cell lack of life, does not require surgical procedure and complements the drift of vitamins inside the area. This creates a healthy surroundings that does not honestly relieve the debilitating ache, however heals the supply of the ache.

Sports Injuries

How many athletes had to upfront say good-bye to the sports activities activities sports region due to injuries sustained while playing their activity? In basketball by myself, famend cagers were relegated to the sideline because of torn cartilage on the knee known as the meniscus. The top notch that traditional medicine has to offer is to control the pain through the use of steroidal injections or thru manner of arthroscopic surgical operation to lessen the damaged tissue.

With stem cellular transplants, sports activities-obtained accidents are given a renewed ray of desire for recovery and healing sans the surgical treatment and with the gain of on the spot deliverance from ache. The opportunity remedy offers patients with osteopathic injuries the capacity to regenerate broken tissues all over again to their real state previous to the damage.

Based at the money owed of skilled osteopathic practitioners, the autologous mode of stem cell harvesting transpires in pretty an entire lot 5 minutes and pain is not frequently skilled besides for a bit stress on the same time because the blood from the bone marrow is extracted thru aspiration. This calls for notably much less time than sporting out guided X-ray epidural injections.

The bone marrow extracted then undergoes apheresis in a centrifuge for 15 minutes. Being spun within the centrifuge gadget motives the unspecialized mesenchymal cells to be separated from the platelets and different components of the blood. After the filtering way, the osteopathic surgeon infuses lower back a affected person's stem cells into the damaged region.

The mesenchymal cells are later transformed into bone tissue, blood vessels, cartilage, ligament, muscle tissues, nerve tissue or tendon. Stem cells feature to heal sports accidents within the following manner:

•Locate the vicinity that needs to be repaired;

•Communicate with the DNA with recognize to an appropriate genetic code of the affected man or woman; and

•Perform the damage restore and regeneration based at the commands communicated thru manner of the genetic code.

Stem cells are able to repair the harm because of the truth the DNA code tells them what they need to be or what they want to reveal out (or specialize) into in order for cell restore to take effect. With such interaction most of the code and the stem cells, the latter are in a position to differentiate based totally definitely at the precise necessities for harm repair. However, it's also your genetic code that dictates how properly a broken tissue can restore itself. Your tissue can only regenerate and characteristic, as well as it had at the exceptional condition of your lifestyles.

Arthritis

Sadly all situations that osteopathic treatment handles are simply painful. Do you have got any idea how painful arthritis is? To factor out some, it's miles pain that:

•Penetrates deep down into the joint and may even damage as some distance as the groin, thighs or buttocks;

•Occurs as quickly because the affected joint is used;

•Worsens due to the reality the day progresses and might limit your work average overall performance;

•One feels in spite of the fact that in reality status, sitting or ascending the steps.

Arthritis pain receives even greater painful due to the fact the swelling gets large. The amazing information, however, is that stem mobile remedy can assist reduce joint infection and the accompanying ache. Stem cells inherently possess anti inflammatory houses. Moreover, they also can sluggish down the method of degeneration inside the affected joint. In addition, it has in recent times been determined that stem cells have the capacity to provide rheumatoid arthritis pain remedy, on account of its efficacy in turning in protein into the broken region.

Research also showed that stem cells do now not just restore focal harm on the joint cartilage, but works to heal the entire joint.

Stem cells are able to do that with the aid of regenerating sufficient cartilage to offer a cushion for the joint and contrary the harm. At the very least, they save you the improvement of the damage at the cartilage, or in first rate phrases, stop the cartilage from typically degenerating.

Tendonitis

As people age, their bodily restore systems decrease in ordinary overall performance. In exclusive phrases, as human beings get older, their our bodies' capability to heal from the damage and tear of exercise, accidents and playing is reduced. Advancing years cause someone's mesenchymal cells to reproduce in lesser portions.

While stem cells can't make humans younger once more, the ones unspecialized cells have the functionality to rejuvenate the healing capability of your very private blood to help you repair broken tissues. The way fictional vampires need blood to hold their young adults and to invigorate their vintage weary our bodies make revel in anyhow on the same time as taken inside the context of stem cells. Yet, thinking about the fact that they do no longer source their sip of revitalizing sparkling blood from special humans, the technique isn't always autologous however allogeneic.

Spinal Stenosis

If you enjoy numbing, very painful burning or tingling in the place in your decrease decrease lower back, your legs and as a ways down as your feet, then you will be laid low with spinal stenosis. This state of affairs takes region with the growth of recent bone, as well as soft tissue on your vertebrae. Such growth effects within the bargain of region on your nerves within the spinal canal. The ache from spinal stenosis usually takes region due to the fact the bone and tissue boom compresses the nerve roots.

Depending on the severity of the pain and the vicinity of the nerve compression, the equal antique interventions for spinal stenosis incorporate non-invasive tactics and remedy. Pain relievers which includes every prescription and non-prescription brands are a fixture of the drug recurring of a affected person suffering from spinal stenosis. In addition, the following drug types can also be prescribed:

•Antidepressants

•Corticosteroids

•Muscle relaxants;

•Non-steroidal anti-inflammatory tablets; and

•Opioids

Autologous stem mobile remedy is an possibility treatment for spinal stenosis. This direction of remedy is powerful for lowering or healing the infection.

Chapter 14: The Future Of Osteopathy And Stem Cell Therapy

The project of osteopathic treatment isn't satisfactory restricted to sports sports activities injuries, sciatica, arthritis, tendonitis or spinal stenosis. There are some of different situations under osteopathic remedy in which stem mobile remedy may be finished which include the subsequent (be aware: this is not an exhaustive listing):

•Bursitis

•Carpal tunnel syndrome

•Cuff syndrome

•Fibromyalgia

•Fracture

•Frozen shoulder

•Hip pain

•Low decrease lower again ache

•Muscle spasm

•Neck and lower back ache

•Neuralgia

•Overuse injuries

•Repetitive stress damage

•Sprain

•Whiplash injuries

•Wrist ache

With the winning quantity of evidence from studies and the outcomes documented from anecdotal data of osteopathic practitioners, the technological statistics of osteopathic medication can fuse with regenerative medicine thru autologous stem cell transplantation and treatment. Stem cells are slowly getting wider publicity and hobby for his or her healing and mobile restore skills.

With stem cells, broken organs, tissues and cells in your body may be replenished and repaired. For example, stem cells will assist in generating new cells in case your pores and skin tears and gets prone. The potential of stem cells to regenerate a broken liver really cannot moreover be wondered.

The remedy is presently to be had even outside the USA. Hundreds, even masses, of sufferers have already been treated by means of using this remedy. Most of them completed exceptional consequences after undergoing simply one session. Patients of chronic Lyme contamination, coronary coronary coronary heart troubles and rheumatoid arthritis showed dramatic enhancements in their fitness after present process the system.

Stem cells are part of everyone. They may additionally moreover additionally however be the surprise healers that clinical technological understanding is searching out for the purpose that the advent of modern-day-day remedy. Stem cells help regenerate tissue. They may be the savior of sufferers reeling from the effects of osteopathic injuries. And preliminary effects located that remedy is felt nearly right away for a few patients at the identical time as others say inner only a do not forget of days.

Chapter 15: Cell Differentiation

The mobile differentiation is a idea from developmental biology describing the method via the use of which cells recognition on a "commonplace" cell. The morphology of a mobile can also alternate significantly all through differentiation, however the genetic cloth remains the same, with a few exceptions.

A mobile able to differentiating into numerous cellular sorts is called pluripotent . These cells are known as stem cells in animals and meristematic cells in better plant life . A cell capable of differentiating into all the cellular types of an organism is called totipotent . In mammals , only the zygote and more youthful embryonic cells are totipotent, at the same time as in plant life, many differentiated cells can become totipotent.

Epithelial Cells

Image of epithelial cells (pores and pores and pores and skin). The cell nucleus is green and the membrane is crimson.

Diagram of a cone (mobile of the attention)

Representation of a conical cellular of the attention, charged with the vision of colors. We see that its morphology may be very one in every of a kind from that of the epithelial cells, because the 2 cells carry out very one in every of a kind functions.

In most multicellular organisms, not all cells are identical. They present crucial variations of their morphology and feature. For instance, the cells composing the pores and skin in human beings are special from the cells composing the internal organs. However, all the one-of-a-type cell sorts are derived from a unmarried fertilized egg cellular and this, via differentiation. Differentiation is a mechanism via which a non-specialized cellular specializes in one of the many cell kinds composing the frame which encompass myocytes (muscle cells), (Liver) cells or neurons (cells of the anxious gadget). The restrict of the differentiation functionality of a cell, that is to say in the route of which cell sorts it is able to evolve, begins very early at some stage in improvement. In people, as in other tribal metazoans, the cells of the embryo are prepared into 3 zones, referred to as embryonic leaflets . Each of the 3 leaflets (endoderm, mesoderm, ectoderm) can differentiate simplest to unique organs. For example, all the cells of the concerned tool come from the ectoderm. During differentiation, a few genes are expressed, Others are repressed. The gadget of differentiation is inherently regulated thank you in particular to the epigenetic cloth of the cells and in particular transcription factors unique to a given mobile lineage an outstanding manner to comprise a cell despite the reality that naive in a pathway of differentiation (embody MyoD and Myf5 for skeletal striated muscle cells). Thus the differentiated cell will specific a selected part of its genome and boom precise structures and accumulate first-rate abilties. The technique of differentiation is inherently regulated thank you specifically to the epigenetic cloth of the cells and specially transcription factors unique to a given cell lineage at the manner to contain a cellular despite the fact that naive in a pathway of differentiation (encompass MyoD and Myf5 for skeletal striated muscle cells). Thus the differentiated cell will explicit a particular part of its genome and increase unique systems and gather excessive satisfactory functions. The technique of differentiation is inherently regulated thanks specifically to the epigenetic material of the cells and specially transcription factors unique to a given cellular lineage so you can comprise a cellular although naive in a pathway of differentiation (consist of MyoD and Myf5 for skeletal striated muscle cells). Thus the differentiated mobile will specific a selected part of its genome and growth precise structures and gather incredible skills.

Differentiation can bring about modifications in lots of additives of the body structure of the cellular: its period, form, polarity, metabolic hobby , sensitivity to tremendous signs and gene expression can all be altered at some stage in differentiation. In cytopathology , the level of cellular differentiation is used as a degree of the improvement of a cancer .

Mammalian cells

Separating mammalian cells into three commands: the cells of the germline , the somatic cells and stem cells . Each of the 10 14 (a hundred thousand billion) cells of the human body has its very very own replica (s) of the genome , apart from fine cells which have out of place their nucleus whilst differentiated, as is the case for purple blood cells . The majority of these cells are diploid , ie they have copies of every chromosome . These cells are called somatic cells. Most of the cells that make up the human body are on this magnificence.

Germ mobile cells are the cells that eventually provide gametes - oocytes and spermatozoa - and are the most effective ones to transmit their genetic cloth to subsequent generations. Stem cells, as an alternative, have the potential to divide a completely big amount of times and redecorate into specialised cells whilst regenerating.

Differentiation at some point of improvement

The improvement begins offevolved while a sperm fertilizes an egg and creates a single mobile which can likely form a whole organism. In the number one days after fertilization, this egg-cellular divides into numerous same cells. In man, about 4 days after fertilization and after numerous mobile cycles, those cells begin to specialize and shape a hole sphere known as blastocyst . This one has a layer of out of doors cells (the peripheral cells or trophectoderm) and a collection of internal cells, referred to as cells of the internal mass. It is those cells so as to shape all of the tissues of the human body. Despite this, They can no longer for my part shape an entire organism; they're referred to as pluripotent. These cells then live frequently decided to offer stem cells that allows you to supply cells of well-defined sorts. For instance, blood stem cells in the bone marrow produce red blood cells , leukocytes and platelets .

Differentiation over the route of life

Stem mobile differentiation is a mechanism that lets in human beings to resume their cells. The basal a part of the pores and skin includes stem cells, which range asymmetrically: a stem mobile gives a pores and skin cell (keratinocyte) and a stem cell. The cellular of the formed pores and pores and skin regularly migrates to the ground of the pores and skin. Thus, our dermis is continuously renewed.

In the same manner, the intestines are protected with small protrusions, the villi. At the lowest of its protrusions is a crypt, which houses a stem cell. The daughter cells of this final one often migrate upwards of the villi. As fast as they are at a positive distance from the bottom of the crypt, they no longer feel the movement of proteins Wnt (which inhibit differentiation). They for that reason become endothelial cells 1 .

Dedifferentiation

It might be noticed, therefore, that due to the fact the cells vary, the type of cellular kinds which they may produce diminishes, sooner or later the call of specialization. However, to a super amount, there are dedifferentiation phenomena via which quite specialized cells can come to be lots a lot less specialized. This type of mechanism stays restrained insofar as epigenetic material (particularly) is irreversibly modified at some degree within the manner of differentiation.

In animals, this phenomenon is uncommon within the herbal u . S ., but we are able to deliver the example of the tail of the triton : after being lessen, the cells of the stump are dedifferentiated, so one can be capable of reform all of the tissues of the tail .

Plant cells

Some cells will differentiate into absorbent hairs (a cellular = an absorbent hairs); Other cells will represent the conductive vessels of sap, cells of the parenchyma ... And so on. These cells are made from meristematic cells of the meristem cauline (stem and leaf) and meristem root (root) 2 . The meristematic cells prevent their proliferation and differentiate definitively after the floral induction and formation of the tissues of the flower. Plant cells may be dedifferentiated cells as pericycle that may reason secondary roots.

Pathology

In certain pathological occasions, cells can also moreover additionally alternate differentiation. This is metaplasia . For instance, under the have an effect on of inhaled smoke from tobacco , the breathing cells of the bronchial mucosa may be transformed into squamous cells .

Moreover, sooner or later of the most cancers technique , the differentiated cells can lose their differentiation and emerge as anaplastic .

In immunohistochemistry, it's miles viable to look at particular proteins of a given histological type, called " differentiation marker ".

The cellular department is the multiplication of any mode cellular . It allows it to divide into severa cells (most often). It is therefore a critical technique inside the living international, considering it's miles important for the regeneration of any organism.

In eukaryotes - characterised mainly by way of way of cells that own a nucleus - there are varieties of mobile division:

The mitosis which lets in most effective asexual ; It lets in regeneration of an organ, and additionally growth.

The meiosis that permits sexual reproduction .

In the prokaryotes , cellular division is finished thru manner of Scissiparity . These cells typically have a unmarried chromosome that replicates before the two chromosomes separate and the relaxation of the mobile divides in flip.

Chapter 16: Reversal Of Department Way

In April 2006 , the Oklahoma Medical Research Foundation International claims to have positioned a device reversing cell division [archive] , that might result in new techniques of combat against most cancers (similarly to new guns preventing the recovery of wounds). This discovery come to be added via Nature mag in its trouble of April thirteen, 2006.

From embryos (embryonic stem cells),

From unfertilized eggs,

From embryonic stem cells changed inside the laboratory,

From a genetically reprogrammed mature cellular,

From a differentiated and mature cellular and then cultivated within the laboratory.

Medical Applications

In medication , animal and human stem cells have been the situation of a superb deal research for the reason that Nineteen Nineties , with the preference of regenerating tissues , or maybe of creating whole tissues , and preferably of reconstructing organs (cell remedy) of The equal way due to the fact the opozones nine , invented through Auguste Lumière . These capability benefits have brought on healing cloning experiments to manipulate the manufacture of those merchandise in large numbers.

The first drug made of stem cells authorised in May 2012 thru the Canadian authorities. This is the Prochymal , a education derived from man or woman stem cells mesenchymal 10 .

Animal stem cellular

An animal stem mobile is an animal cellular characterized through its capability to generate specialized cells through differentiating and by using being capable of multiply identically (self-renewal).

Animal stem cells and mainly human stem cells are the trouble of a remarkable deal studies, particularly in treatment, so one can regenerating tissues or even growing tissues and organs . The use of stem cells in medication poses moral problems .

Metazozoic organisms are the result of embryogenesis , this is, they're fashioned from a single cell (the zygote), which in flip develops into an entire character. Some cells of an organism end result straight away from this embryogenesis, but many others are constantly renewed: Stem cells descend from those embryonic mobile populations and play a cell "reservoir" position thru the usage of changing the cells out of place each day or in some unspecified time in the future of accidents.

We distinguish severa kinds of stem cells consistent with their ability of differentiation:

The cells totipotent stem : or fertilized egg cells from the primary divisions of the egg to the morula degree (2 to eight cells). These cells are the best ones to allow the improvement of a whole individual supplied that they stay in-vivo to allow an embryo orientation now not viable in-vitro . Etymologically totipotence manner "all power" indicating that theoretically those cells may be differentiated in any mobile form of the organism that they have got to result in shape (epithelial, neuronal, hepatic cells ...).

The pluripotent stem cells which might be part of ES cells (embryonic stem): ES cells can not produce an entire organism, however can differentiate into cells derived from any of the three germ layers , which incorporates germ cells. They by myself cannot result in the advent of a complete person. They originate from the internal cellular mass of the blastocyst (at the diploma of forty cells) even as the placenta which nourishes the embryo and protects it from any rejection thru the immune machine is produced thru the outer mobile layer (or trophectoderm). Reproductive cloning from ES cells isn't always feasible.

The multipotent stem cells : gift in the embryo or within the person organism, they'll be the begin of severa styles of differentiated cells. Multipotent stem cells can deliver upward thrust to numerous kinds of cells , however they are already engaged in a superb direction. They are said to be particular cells. Their possibilities are therefore extra confined than the ones of ES cells. The hematopoietic cells in mammals, for example, produce crimson blood cells , the platelets , the T cells or B , the macrophages , however they can not supply the cells muscle .

The mobile colony forming can produce simplest one cell type (on the equal time as autorenouvelant) and are decided in some organs such as pores and pores and skin, liver, intestinal mucosa or testes. Some organs, consisting of the coronary coronary heart and the pancreas, do no longer include stem cells and for this reason haven't any possibility of regeneration inside the occasion of harm.

The term "stem cell" have to no longer be used to consult cells which, even though capable of differentiating into one or greater cell kinds, cannot self-renew in a truly countless way. This is the case of " progenitor cells ", which are very not unusual within the body and feature constrained functionality for branch.

Role

Some tissues have a strong mobile renewal and a very crucial quantity of stem cells. In mammals , they regenerate the subsequent cells:

The epidermal cells which give safety from the out of doors, and which often damage aside;

The cells of the intestinal crypts , which undergo the conditions crucial for the hydrolysis of food , and which must therefore be hastily renewed;

Blood cells (erythrocytes , platelets , leukocytes , and so on.) all of that are crafted from a single shape of stem cellular, the hematopoietic stem cells contained within the bone marrow ;

In male people, the spermatogonies which create huge portions of spermatozoa in the course of the way known as spermatogenesis ;

the neural stem cells .

These cells can not be discerned experimentally in a cell populace, both with markers or morphologically, while you do not forget that they may be undifferentiated and consequently do no longer have any differentiation markers. They are in fact defined thru their function and no longer their shape. Their discovery can therefore nice be completed in the route of experimentation.

Stem cells are particularly harassed inside the occasion of a stunning alternate inside the environment.

Since 2007, regular cells (fibroblasts) may be converted into stem cells with the useful useful resource of genetic engineering thru incorporating numerous genes . The manipulation remains tough, however, with a immoderate failure fee 1 .

Types of stem cells

totipotency

The totipotency is in biology , the assets of a cell to distinguish into any specialised cellular shape itself and forming a living multicellular 1 . Totipotence is in opposition to pluripotency , multipotence , and unipotence .

Biology

In mammals , the fertilized oocyte and each daughter cell (blastomer) of the embryo at a few degree within the first actual cell divisions are the most effective real totipotent cells. For man, the totipotency of the blastomers of the embryo with or 4 cells is attested thru the births of monozygotic twins and monozygotic quadruplets 2 , three .

While pluripotency , multipotency , and unipotence are stem cellular homes , in mammals the totipotency belongings does now not comprise stem cells. Indeed, this belongings is misplaced very suddenly in vivo at some point of the primary cell divisions. The totipotent cells do no longer consequently revolve however differentiate themselves inside the course of the trophoblast and the internal cell mass of the blastocyst . While protocols exist to hold in way of life in vitro cells pluripotent (eg embryonic stem cells), no such n '

The plant cells have the distinction of being generally totipotent.

The pluripotency is the functionality of great cells to distinguish in: cells of the three germ layers (ectoderm , mesoderm and endoderm), cells of the trophectoderm 1 or germ cells .

Embryonic stem cell

A colony of the HD90 human embryonic stem cell line derived from the Montpellier CHRU.

This distinguishes them from unipotent , multipotent , and totipotent stem cells (which could differentiate in any form of specialised mobile).

The pluripotent cells can't produce an entire organism as they proliferate and differentiate in an anarchic manner. In vitro , they spontaneously shape embryoid our bodies.

The 3 sorts of pluripotent stem cells described thus far in a reproducible manner are:

the embryonic stem cells (ESC);

the brought on pluripotent stem cells (IPS);

MUSE cells (for Multiline-differentiating Stress Enduring 2) decided within the pores and pores and skin and bone marrow (1 cell out of 5000) of adults.

Biology

Obtaining

Embryonic stem cells are remoted in vitro from the internal cellular mass of the blastocyst (on the five or 6 th day for human embryogenesis). Induced pluripotent stem cells (iPS) are derived from the reprogramming of adult somatic cells into pluripotent cells with the useful resource of overexpression of nice transcription elements.

Markers

There are some of alternatively precise floor markers of pluripotent cells along with SSEA-3, SSEA-4, TRA-1-60, TRA-1-eighty one (in people) or SSEA-1 (in mice). Alkaline phosphatase interest is likewise an remarkable marker for undifferentiated pluripotent stem cells. Recently, new markers have been recognized the usage of DNA chips: CD24, SEMA6A, FDZ7 3 .

Transcription Factors

Pluripotent stem cells are characterised by using sturdy expression of the transcription elements OCT4 / POU5F1 , NANOG 4 , five and SOX2. These three transcription factors are the "middle transcriptional regulatory circuitry" 6 .

Growth Factors

Leukemia Inhibitory Factor (LIF) 7 and Bone morphogenetic proteins (BMPs) eight are required for the cultivation of mouse pluripotent stem cells. In assessment, bFGF is wanted inside the way of life medium to keep the pluripotency of human pluripotent stem cells, while BMPs spark off a lack of their pluripotency (differentiation).

Applications

The have a look at of pluripotent stem cells allows a better expertise of the early tiers of embryonic improvement. Genetically bizarre pluripotent stem mobile lines are a version for the have a observe of unusual genetic illnesses. They also can check new pills.

Medicine

Pluripotent stem cells might be a very interesting source of cells for regenerative remedy : infinite in vitro amplification, possibility of differentiation in any cell kind. They open the way for cell treatment of the sickness Parkinson , of diabetes , of myocardial infarction , and so on. The essential obstacles to those packages at present are: the formation of teratomas due to their intense and unregulated proliferation while injected inside the undifferentiated u . S . A ., hassle in acquiring mature cells capable of repairing the broken organ and The trouble of immunological compatibility, Major interest of iPS (delivered on pluripotent stem cells). Pluripotent stem cells may additionally moreover have an interest in gene therapy .

Ethics and Legislation

Research on pluripotent stem cells has lagged in the back of in France due to the ban on studies on human embryos because the Bioethics Act of 1994. In 2004, this law become revised 9 : the ban is maintained with notwithstanding the reality that The opportunity for research organizations to reap a derogation for 5 years beneath positive situations: "studies can be legal on embryos and embryonic cells wherein they may be likely to result in primary recuperation advances and supplied that they can not Be pursued through an opportunity technique of comparable performance, in the nation of clinical records '. Embryos should had been designed as a part of an MPA, Agreement of every mother and father. The complete is supervised with the useful resource of the Agency of the biomedicine. The discovery of iPS permits researchers to behavior studies on pluripotent stem cells without strolling on the embryo. However, iPS and embryonic stem cells are similar however no longer identical: it stays to represent those versions and recognize their impact for regenerative remedy.

Neural stem cells

The neural stem cells are cell lines - multipotent and able to self-renew one -

whose capability for differentiation is confined to neural cell types, consisting of:

Neuron 2 ,

Astrocyte 2 and

Oligodendrocyte 2 .

During embryogenesis , those neural stem cells are located in the ventricular quarter of the neural tube .

They generate all of the mobile kinds needed with the useful resource of using the substantial stressful system (besides for microglia), with the aid of way of the usage of a machine known as neurogenesis .

Unlike what have end up concept at the begin of the xx th century, neurogenesis does no longer get up handiest within the course of embryonic improvement and to formative years, but maintains to conform physiologically in the course of person existence. An orphan nuclear receptor , the "TLX" plays a function inside the endurance and proliferation of person neural stem cells via manner of suppressing their differentiation to a glial phenotype 3 .

Presence in character mammals

In mammals excellent areas of the mind are regarded to maintain man or woman neural stem cells four :

The sub-ventricular area (ZSV) which borders the lateral ventricles ; The neuronal precursors produced in this sub-ventricular region then migrate to the olfactory bulb , wherein they differentiate into mature neurons;

And the toothed gyrus of the hippocampus (or "GD"), which plays an vital characteristic in memorization .

In laboratory rats , research have demonstrated the feature of these neural stem cells in a few techniques of memory 5 , melancholy 6 , and smell (smell memorization 7).

There can also be neural stem cells inside the individual peripheral worried device .

Challenges

Many medical hopes relaxation at the presence of neural stem cells in strains glial and neuronal grownup, notably the combat in opposition to neurodegenerative diseases (Alzheimer's sickness and Parkinson's disease) or that facilitate neuronal regeneration After a lesion (neurological harm or ischemia).

Characterization

Neural stem cells are in particular studied in vitro , the usage of a way of existence device called test of neurospheres that have become advanced with the aid of way of way of Reynolds and Weiss 8 . The cells are extracted from a potentially neurogenic place (or purified via way of floor markers or genetic changes) and are cultured within the presence of boom elements (eg EGF, FGF). The cells then form:

Or neurospheres, a round-formed cell grouping - indicating a first rate neurogenic capacity,

Or a mat of differentiated cells - indicating a probable absence of stem cells within the test location.

The neurospheres shaped include heterogeneous populations, which incorporates slowly dividing neural stem cells (1 to 5%) and the large majority of speedy-dividing nestin-terrific progenitor cells eight , 9 , 10 . The overall range of those progenitors (with sluggish and fast divisions) determines the scale of the neurospheres. Thus, a disparity within the size of the spheres among populations of different origins also can replicate versions in proliferation, survival and / or differentiation developments. For instance, the deletion of ß1- integrin does no longer save you the formation of neurospheres however considerably reduces their duration:

The houses of self - renewal and multipotence of neural stem cells had been described specifically via using this neurosphere system:

The ability for self-renewal corresponds to the functionality to form secondary neurospheres from cells remoted from the number one neurospheres.

In a 2d step, the withdrawal of increase elements and the following differentiation of the cells makes it viable to understand the multipotence of those neurospheres. The presence of neurons and glial cells demonstrates that they will be crafted from neural stem cells.

This characterization in vitro has obstacles, however, 12 and it is lots extra difficult to demonstrate the character stem cellular in vivo four .

The epidermal increase trouble (EGF) and fibroblast increase thing (FGF) are boom elements for stem and progenitor cells in vitro , however different factors synthesized with the useful resource of the stem and progenitor cells in manner of life are desired for his or her growth thirteen .

Embryonic Origin

The entire frightened system derives from the neuroectoderm, one of the embryonic layers that is awesome at some stage in neurulation. It office work a pseudo-stratified neuroepithelium that surrounds the destiny cerebral ventricles while the neural tube closes . The neural stem cells are at the bottom of this leaflet, at the threshold of the ventricles. They start to have very early neurogenic interest, in the direction of E9-10 (embryonic improvement day) in mice, and acquire radial glial residences : They are in touch with the apical and ventricular surfaces however their mobile frame stays in the vicinity Ventricular valve that bounds the ventricles. During the thickening of the cortex, the apical prolongation lengthens and gives the cellular this radial polarized morphology. Radial glia is characterised thru the expression of various molecular markers, a number of which can be unique to the glial lineage (eg GLAST, BLBP, Nestin, Vimentin, RC1 RC2 and every so often even GFAP). It is on the equal time because the neuroepithelial cells exchange into radial glia that they grow to be neurogenic and pass from a symmetrical proliferative branch to a neurogenic uneven department. The cells of the radial glia generate the majority of neuronal and glial cells of the applicable worried machine 14 . Is whilst the neuroepithelial cells become radial glia as they come to be neurogenic and skip from a proliferative symmetric branch to an uneven neurogenic department. The cells of the radial glia generate the bulk of the neuronal and glial cells of the number one fearful device 14 . Is while the neuroepithelial cells alternate into radial glia as they grow to be neurogenic and bypass from a proliferative symmetric department to an choppy neurogenic division. The cells of the radial glia generate the majority of the neuronal and glial cells of the crucial apprehensive machine 14 .

One of the peculiarities of neurogenic department is interkinetic nuclear migration . The mobile frame of the cell actions alongside the radial extension as a characteristic of the phase of the cellular cycle. The reasons for this phenomenon aren't but stated however might be associated with the law of publicity to differentiation or proliferation factors (eg Notch pathway) or to optimization of area in the ventricular area.

Radial glia can at once generate differentiated cells destined to turn out to be neurons: neuroblasts. These neuroblasts migrate alongside the radial extension of the cellular that generated them. They then fill the cortex thru differentiating into a postmitotic neuron. Radial glia can also generate an intermediate proliferative precursor an excellent way to distinguish into neurons after numerous cycles of speedy department, because of this growing the wide variety of neurons generated.

All neurogenic interest is tightly controlled to generate the ok type of neurons. The neurogenic after which gliogenic interest is probably regulated ordinary with time and space to generate the right form of neuron in the right place. This control might be completed by morphogenic sports activities activities and transcriptional programs precise to every part of the mind and the spinal twine. One of the splendid troubles is whether each radial glia inside the ability to generate all the various glial and neuronal populations or there are specific populations of radial glia, every with a constrained capability 14 .

Chapter 17: Discovery Of Individual Neural Stem Cells

As early as 1960 , the formation of new neurons in a part of the hippocampus turn out to be suspected in the course of postnatal existence and in teenagers , notably through Altman and Das in 1965. Around 1970 , André Gernez and his collaborators assert that neurogenesis Continues to exist after transport . Nevertheless, the significance of these consequences have end up now not exploited and the situation remained arguable for nearly many years.

In 1989 , the crew of Sally Temple defined the lifestyles of multipotent progenitor cells and can self-renew within the subventricular region of the thoughts of mice 16 . In 1992 , Reynolds and Weiss had been the primary to isolate neural stem and progenitor cells from a dissection "coarse" the striatum containing the subventricular place of person mouse brains eight . In the identical year, the team of Constance Cepko and Evan Y. Snyder were the number one to isolate multipotent cells from the mouse cerebellum, transfect them with the v-myc oncogen and reimplant them in a new baby thoughts 17 . His art work opened the door to the possibility of producing new neural cells inside the thoughts from stem cells. In 1998 , Elizabeth Gould of Princeton University confirmed neurogenesis in a specific a part of the hippocampus of the person monkey . This identical phenomenon is located thru Freg Gage institution at the Salk Institute in California within the character 18 . Since then, stem and progenitor cells have been remoted from distinct components of the imperative anxious system, together with the spinal twine , in extremely good species, which includes people 19 , 20 , 21 .

Embryonic stem cells

An embryonic stem cellular is a pluripotent stem cell derived from the internal mobile mass of a preimplantation embryo on the blastocyst degree 1 , 2 . A human embryo reaches the blastocyst diploma 4 to five days after fertilization and includes a cluster of 50 to 100 and fifty cells (inner mobile mass and trophectoderm). Insulation of the inner cellular mass requires the blastocyst to be destroyed.

Embryonic stem cells are a near ideal deliver for transplants and tissue engineering. The capability of a stem mobile to generate all of the body's cells makes it a key device for studies on human ailments, in particular genetic, or for in vitro trying out of drug toxicology.

However, embryonic stem cell isolation break an moral hassle that wishes particular to set up whether or not an embryo at the preimplantation diploma has the identical jail and moral rights as a person or girls greater evolved three , 4 . Conversely , does no longer using those existence-saving cells be proper, whilst the superimposed preimplantation embryos will now not be used and consequently destroyed ? There isn't any consensus and law on the procurement and use of those embryonic stem cells varies from usa to u.S..

One of the alternatives to embryonic stem cells is to use brought on pluripotent stem cells which is probably generated from differentiated cells (eg, pores and pores and pores and skin cells) and consequently do not present the identical moral catch 22 situation.

Embryonic stem cells have been diagnosed in 1981 inside the mice by means of the usage of way of Martin Evans , Kaufman and Martin 5 , 6 , and in 1998 at the man or woman thru the groups of James Alexander Thomson , Joseph Istkovitz-Eldor and Benjamin Reubinoff 7 , eight .

Biology

The embryonic stem cells (ESC) remoted in vitro in the inner cell mass of the blastocyst (embryo at the 5 or 6 th days in humans).

The inner cellular mass has the property of giving start in vivo to the 3 embryonic layers (endoderm , mesoderm and ectoderm) at the beginning vicinity of all the tissues of an person man or woman.

This in vivo pluripotency property can be very short out of vicinity at some point of embryonic development.

In 1981, the mice were capable of maintain in vitro within the pluripotent america of the us of cells of the cellular mass of a blastocyst and to steer them to proliferate indefinitely: it modified into the begin of embryonic stem cells . These cellular lines are genetically regular, in evaluation to malignant tumor strains. At any time, even after severa years of tradition, it's far even though possible to distinguish these cells in any cell type. If embryonic stem cells of mice - which encompass whether or no longer or no longer they were genetically changed - are injected proper proper into a blastocyst of mice, they make a contribution to all tissues of the younger mouse, at the side of the germline.

In 1998 , these results had been reproduced in people via the use of the derivation of the primary human embryonic stem mobile lines. The pluripotency belongings is attested in vivo by means of the formation of teratomas comprising differentiated tissues from the three embryonic layers, after injection with SCID mice. In vitro , using specific way of life situations results in the differentiation of human embryonic stem cells to mature cell sorts alongside side neurons , cardiomyocytes , hepatocytes , hematopoietic cells and so forth.

In 2009/2010, it changed into proven - in mice - that neurons derived from CSE grafted into the visible cortex associated with different neurons preserving off regions of the thoughts of which they'll be no longer precise with out, as an instance, connecting to neurons Of the spinal wire. While CSE grafted into the marrow established it with out connecting to the seen cortex or to extraordinary organs 9 . If that is additionally actual in individual mice, a contemporary pathway of bodily regeneration of cerebral nerve harm might be opened.

Legal

The law n ° 2004-800 of 6 August 2004 on bioethics [archive] amending the 1994, gives for exemption for a duration of 5 years to behavior research on human embryos, "the studies may be legal on Embryonic cells and embryonic cells wherein they're capable of most important to essential healing advances and provided they cannot be pursued with the useful resource of an possibility technique of comparable efficacy in the country of clinical statistics '. This studies is supervised with the resource of the Agence de la Biomédecine .

Furthermore, the 18 October 2011, The Court of Justice of the European Union rejected, inside the name of Art. 6 of the Directive at the Patentability of Biotechnological Inventions , a patent referring to cells acquired from human embryonic stem cells.

Cell remedy

In the context of "biotherapies", mobile remedy objectives to deal with an organ or an organism by means of manner of supplying cells, more frequently than now not obtained from stem cells , to update or complement failing cells. The range of pathologies targeted via this biotherapy is critical: Alzheimer's , Parkinson's , diabetes , leukemia , and so forth. It may additionally facilitate transplants or organ regeneration. This involves awakening the mobile's coding features, directing them within the direction of the popular intention, and permitting the cells or organs produced to be fashionable into the recipient organism (which can also be a donor).

Cell treatment is a new treatment made possible through the use of the usage of present day successes in human stem mobile cultures . The modern day development of cells iPSCs moreover offers any cellular type from an grownup mobile with out the usage of healing cloning and the destruction of an embryo.

Routine use

Every day, sufferers are treated and cured with the aid of way of mobile treatment for blood cancers in hematology. Hematologic stem cells from a donor or from the patient are recovered, cultured and then reinjected to colonize the bone marrow . This is respectively referred to as allografts and autografts of stem cells . Another clinical software of cell remedy is the transplantation of islets of Langerhans .

Successful experiments

In 2011, the corporation of Professor Luc Douay succeeded the number one autotransfusion of a pellet of pink blood cells acquired thru subculture and differentiation of stem cells 2 .

In 2012, a ebook of the Lancet suggested the protection of autologous intracoronary injection of cells derived from the cardiosphère and regeneration of damaged tissue after myocardial infarction three .

By 2014 A RIKEN group , led by way of researcher Masayo Takahashi, performs retinal mobile transplantation on a 70-yr-antique affected man or woman with age-associated macular degeneration (AMD) inside the international's first medical study, Using CSPi four .

Experiments in development

In France

Biotherapies are experimented through way of severa research groups:

A remedy of high quality cardiac insufficiencies through muscle cells taken from the thigh is examined: Magic test 5 ;

A treatment of ischemia of the decrease limbs with the aid of the use of injection of the affected individual 's bone marrow cells is tested by the usage of the European clinic Georges - Pompidou , with the Necker health facility , those of the Timone and Sainte - Marguerite in Marseille, By Joseph Emmerich 6 . Similar experiments are finished on the Grenoble 7 , Reims and Amiens university hospitals ;

the attempts to deal with Huntington's disease with the resource of transplantation of neural stem cells eight , 9 .

Worldwide

In October 2010, an American commercial company company, Geron Corporation have become granted the right to check embryonic stem cells from a affected man or woman with harm to the spinal wire 10 .

History of stem cell studies

Nevertheless, many questions stay, and to this point, only some pathologies can be handled through cell-based totally techniques. The shortage or inaccessibility of person stem cells, the absence of markers to physical become aware of and purify them, and our fantastically limited information of the fundamental mechanisms underlying their self-renewal are all motives that restriction their use in clinics.

Embryonic stem cells, with the aid in their superb capability for proliferation and differentiation, appear as an possibility to character stem cells, however their manipulation in human beings is on the inspiration of a huge debate in our societies for reasons each ethical , Legal and religious.

Lastly, the use of stem cells in a recovery framework requires that the experimental conditions allowing their amplification in vitro and in vivo be defined with awesome precision, to verify their useful integration in the injured tissue and to make certain the absence Of tumorigenic potential.

All the ones points is probably blanketed on this paintings a extraordinary manner to permit readers to get in touch with the contemporary problems and barriers of stem cellular biology.

In the number one a part of our have a look at we will find out the one-of-a-type candidate stem cells for cell treatment.

In the second one element we're able to have a test the tremendous manipulations and experiments carried out on those cells: amplification, purification ...

In the 0.33 a part of our art work, we are able to test the numerous fields of studies in cellular treatment, focusing greater on cell treatment applied in dermatology, for neurodegenerative ailments which encompass Parkinson's sickness, kind I diabetes and coronary coronary heart sickness.

And ultimately, we are in a position to talk about worldwide law and the ethical and religious revival of stem mobile manipulation.

Chapter 18: Cell And Natural Development

Recall of organic development in man.

Embryology is the study of the embryonic development of the egg mobile to an impartial individual.

The embryo improvement includes various milestones that can be characterised via way of the differentiation and specialization of most of the cells that make up the embryo. [14]

Fertilization :

12 hours after ovulation, the oocyte while fertilized becomes a zygote; The approach of fertilization takes location within the distal 1/three of the fallopian tube and lasts about 24 hours. The date of fertilization is counted because the first day of human ontogenesis or embryonic development.

Fertilization refers to the fusion of two gametes of different sexes into an egg cell at the starting place of a latest person. The egg cellular is therefore totipotent .

Fertilization offers three devices of crucial sports that may be decided in all organisms:

The unique recognition of gametes which ensures the specificity of fertilization;

Activation of the ovum with the aid of the spermatozoa which triggers a tough and fast of programmed metabolic events;

The fusion of parental genomes, a prelude to the division of the egg and the development of a ultra-modern diploid being. [14,15]

The egg mobile will then divide, that is the phenomenon of segmentation .

The segmentation :

24 hours after fertilization, the zygote starts offevolved to go through segmentation, ie a sequence of mitosis resulting inside the

formation of , 4.Eight daughter cells or blastomers ; (Figure 2)

The first branch is vertical, we then benefit blastomers, the second one branch is also vertical but in a perpendicular plane

to the number one. At this diploma, the embryo consists of 4 cells, if each of these cells is isolated and reimplanted proper right into a uterus you in all likelihood can gather a whole person , the cells are totipotent .

The 1/three branch is executed in a horizontal plane, as a give up end result 8 cells are obtained. This branch separates the animal pole as a manner to supply the embryo in itself and the vegetative pole which gives the embryonic annexes. The cells, from this division come to be pluripotent .

The cell cycles will spread suddenly and are synchronous on the begin and then will become asynchronous thereafter: one obtains a morula . In people the morula level is reached four days after fertilization. At this degree the cells are pluripotent, so they're in a position to distinguish into any cellular kinds that make up the organism, but they're capable of not supply the embryonic annexes. The embryo all through this phase maintains the identical diameter and length, most effective the sort of cells will growth.

In the blastula diploma (sixteen to 64 cells), a hollow space seems: it is the blastocele . The cells of the inner mass of the blastula are continuously pluripotent .

They constantly have the functionality to differentiate into cells that make up the three embryonic layers. [14,16,17]

The embryo then enters a step which objectives to vicinity the three leaflets in place: it's miles the diploma of gastrulation

Gastrulation:

Gastrulation is a dynamic section. During this second section of embryonic development, a fixed of coordinated mobile actions: morphogenic moves, redistribute the blastomer format of the blastula and divide it into 3 layers in people: an ectoblast outer leaflet , a median leaflet The mesoblast and an internal leaflet the endoblast from which the organs of the embryo and then the person are constructed; (Figure 3) [14]

At this diploma the cells are multipotent , they will be engaged in a leaflet. It is in the course of the gastrulation that we witness the installing region of the blastopore which is a gap in the blastula. It is from this blastopore, this is an organizing center, that the morphogenic moves will take location.

Morphogenic actions involve three styles of mechanisms:

The invagination of superficial territories inside the embryo

The winding of superficial territories that replicate on themselves via sliding on a community of molecules.

The extension of a superficial territory which spreads in a sheet at the floor of the embryo implying a mobile multiplication, a cell rearrangement of a multi - layered layer which turns into unistratified. [19]

During those great stages, the cells lose their potential for differentiation, this being carried out thru the dedication method.

The determination :

More or an lousy lot less early in development, a splendid variety of traits of the embryo and the character are received definitively and could play an crucial role within the next organogenesis. They are essentially spatial landmarks and the future of cells. This willpower step is not a really defined step in a few unspecified time inside the destiny of embryogenesis because it starts after fertilization and maintains at some point of embryonic improvement.

Cellularly, dedication does not bring about any seen morphological adjustments that might prepare for his or her next differentiation. They sincerely undergo a restriction in their ability for differentiation which becomes restrained to a single direction wherein they may be henceforth engaged. [14,20]

All these mechanisms bring about the status quo of new cell interactions that put together the embryo for the organogenesis phase .

Organogenesis:

The formation of organs takes area progressively subsequently of the improvement of the embryo. It requires a exquisite coordination in the differentiation and scheduling of the tissues involved in their production. This coordination is ensured by way of manner of a series of interactions among cellular groups: a difficult and speedy of cells emits a sign that motives the expression of certain genes in each other commercial enterprise employer of cells and lets in their differentiation in a selected pathway.

The cell electricity of will of the presumptive territories will result in the formation of the organs and their mobile differentiation. The majority of the cells within the route of this degree lose their potentiality, they come to be unipotent , this is to mention that they might differentiate in pleasant one cellular kind.

However, some cells in addition to in adults stay undifferentiated cells, as is the case for embryonic germ cells which can be pluripotent cells.

During this degree the embryonic annexes are set up area allowing the development of the embryo, they'll provide the destiny placenta and the umbilical cord . Stem cells are located in the ones structures, they may be pluripotent or multipotent.

Organogenesis is connected to morphogenesis, that is to mention, the modeling of the frame of the embryo. [14,21]

At 6 weeks, all the organs are fashioned, they need to now amplify to come to be sincerely beneficial, then no longer communicate of embryos but of fetuses.

Stem cells

The cell is the number one unit of living beings. Its duration is a few hundredths of a millimeter. The cellular has a nucleus (except for the Prokaryotes, and some unique cells which encompass red blood cells), it's far surrounded with the aid of way of cytoplasm. The nucleus includes maximum of the genetic facts. In the cytoplasm most of the biochemical reactions necessary for the life of the cell (molecule synthesis and energy transformation) take vicinity.

A man or ladies is product of about one hundred thousand billion cells, belonging to about hundred differing types, maximum of which not divide, at the same time as about twenty million cells of our organism divide to preserve normal the huge fashion of cells (Replacement of the cells disappearing via developing vintage or by way of manner of lesion), those cells are referred to as; Stem cells [22]

All stem cells aren't equivalent, as defined inside the preceding financial ruin. Four categories of stem cells can be distinguished in line with the variety of the cellular types to which they could supply start. However, researchers in mobile remedy do now not hobby On pluripotent and multipotent stem cells. [23]

Totipotent stem cells : present inside the first 4 days of the embryo, they may be the most effective ones to permit the development of a whole organism.

Pluripotent stem cells : or moreover referred to as embryonic stem cells : gift from the 5th to the seventh day following fertilization, they are capable of offer upward push to extra than 2 hundred one in all a type sorts of tissue;

Multipotent stem cells : man or woman and fetal stem cells can provide upward thrust to numerous varieties of differentiated cells.

Unipotent stem cells : they generate only differentiated cells of a unmarried tissue kind and hold sure capacities for self-renewal and proliferation. [23]

One of the peculiarities of neurogenic department is interkinetic nuclear migration . The cell frame of the mobile movements along the radial extension as a function of the segment of the cellular cycle. The motives for this phenomenon are not however diagnosed but may be related to the regulation of exposure to differentiation or proliferation elements (eg Notch pathway) or to optimization of space inside the ventricular place.

Radial glia can at once generate differentiated cells destined to emerge as neurons: neuroblasts. These neuroblasts migrate alongside the radial extension of the mobile that generated them. They then fill the cortex with the useful resource of differentiating into a postmitotic neuron. Radial glia can also generate an intermediate proliferative precursor in case you want to distinguish into neurons after numerous cycles of rapid division, therefore developing the sort of neurons generated.

All neurogenic hobby is tightly controlled to generate the ok significant kind of neurons. The neurogenic and then gliogenic interest is probably regulated everyday with time and region to generate the proper shape of neuron within the right area.

www.ingramcontent.com/pod-product-compliance
Lightning Source LLC
Chambersburg PA
CBHW071223210326
41597CB00016B/1919